The Crochet Lady
Copyright © 2020 Albert Quintana

This book is a work of fiction. People, places, events, and situations are the products of the authors imagination. Any resemblance to persons, living or dead, or historical events, are purely coincidental.

All rights reserved. No part of this book may be reproduced, distributed, or transmitted in any form or by any means, including photocopying, recording, or other electronic or mechanical methods, without prior written permission from the publisher or author, except in the case of brief quotations embodied in critical reviews and certain other noncommercial uses permitted by copyright law.

Library of Congress Control Number	2020918711
Paperback	978-1-63626-036-5
Hardcover	978-1-63626-037-2
eBook	978-1-63626-038-9

Printed in the United States of America

Harrison Street, Hoboken
New Jersey, 07030
www.paperchase-solution.com
+1-800-850-2688

The Crochet Lady

Albert Quintana

Foremost, I give a big thank you to all the supporters of The Crochet Lady.

To my editors,
my beautiful wife Louise Quintana,
my great friend Dolores Bargas,
and cover design to the most creative Carlos Mora.

Dedication to Alonzo Sisneros, son of *The Crochet Lady*

Table of Contents

CHAPTER ONE **Tender Loving Care** 1

CHAPTER TWO **L'Ojito, Colorado 1889** 7

CHAPTER THREE **A Dilemma** 13

CHAPTER FOUR **Mount Blanca 1928** 17

CHAPTER FIVE **The Lion's Gate** 23

CHAPTER SIX **New Mexico State Orphanage** 27

CHAPTER SEVEN **100 days** 37

CHAPTER EIGHT **The Demise** 41

CHAPTER NINE **The Journey to Chama** 49

CHAPTER TEN **A Proposal** 61

CHAPTER ELEVEN **King Henry the Eighth** 63

CHAPTER TWELVE **The Rifleman** 67

CHAPTER THIRTEEN **Search for Gold** 71

CHAPTER FOURTEEN **My Best Friend Alvie** 81

CHAPTER FIFTEEN **The Shepherd** 85

CHAPTER SIXTEEN **Life of Boredom** 87

CHAPTER SEVENTEEN **Once Lost** 93

CHAPTER EIGHTEEN **Capulin, Colorado** 97

CHAPTER NINETEEN **Headaches, Neck Pains, Falls** 99

CHAPTER TWENTY **Love at First Sight** 101

CHAPTER TWENTY-ONE **White Mountain** 105

CHAPTER TWENTY-TWO **The Conversation** 109

CHAPTER TWENTY-THREE **The Celebration of the Virgen de Guadalupe** 113

CHAPTER TWENTY-FOUR **The Letter** 117

CHAPTER TWENTY-FIVE **The Visitation** 123

CHAPTER TWENTY-SIX **The Rendezvous** 127

CHAPTER TWENTY-SEVEN **Crossing the Bridge** 133

CHAPTER TWENTY-EIGHT **On the Mountain Top** 137

An Introduction

Life is a blur…Its duration is minimal to say the least. For some people life is short, while for others live a long time. You never know when the end will come. One day you recognize and acknowledge that you exist on this earth and in a moment you're gone.

The earth from the moon looks like a big blue marble floating into the cosmos. This blue marble is alive and breathes like any living creature. Its movement is deliberate. Scientists, people of great study make educated mathematical calculations on its movement. Their findings are recorded and used as resources for further study. But, this big marble is still a mystery.

The earth thrives on constant movement. Its direction is perpetual and never changes course with its revolutions and rotations. But there's a constant change on its surface. There are tornadoes, earthquakes and hurricanes which create natural changes to this blue marble where we coexist. Humans create changes also when they excavate and build, tear down and rebuild. The most notable change to our surface comes from mining that extract fossil fuels to keep up with demand of energy to fuel cars, warm and cool homes and factories that manufacture steel to build sky scrapers and much more.

But as humans we recognize an existence on this blue marble. We live in an environment and survive the earth's challenges and in the next moment we exit. But this Mother Earth continues to spin. Our remains stay on this earth, but what happens to our spirits, our souls?

Experiencing death can be very traumatic to our being. We have ties and bonds to others and when they leave we are crushed emotionally. We cannot get away from seeing that exit, for it is inevitable.

Many people believe that there is life after death while others believe that "this is it." Do angels take us to the after world called heaven or hell? Hindus believe in reincarnation; where people, plants, animals and humans come back to this earth as plants, animals or other humans, but into another caste or class of Hindus.

Christians believe in a Heaven or a Hell. The good go to Heaven, the evil to Hell. Buddhists believe similarly to Hindu's reincarnation but a focus on Karma. Karma, meaning, what you do in life will reflect on your of life outcome. Treat people good and good things come back to you and vise-versa will reflect how bad your outcome will be.

The Crochet Lady is a proper example of life and death. *The Crochet Lady* through its pages, reflected only love and caring for her family. Her final exit was a hard painful death. For all of the worst ailments a person could ever have, bad heart, loss of hearing, excruciating arthritic pain, and the loneliness she had by not being able to communicate her feelings. It just seemed like it wasn't right or fair for a sick, wonderful, grand lady to leave this world in such a manner.

Her story is beautiful. It is a story of love and hope that exemplifies true dependence on God to guide her life. From the beginning the odds were against her having a great life. However, the love her mother gave her was a true example to how *The Crochet Lady* was to treat others. *The Crochet Lady* is an example to my life. She is a legend in my eyes. In her last days she asked me to write her story. Here is her story. She is *The Crochet Lady*. The promise has been fulfilled!

CHAPTER ONE

Tender Loving Care

History is like the wind.

The Crochet Lady waits patiently for the long day to begin. She sits on her favorite chair by the window, taking in the morning sun. The rays hit her face. She feels the warmth from the bright sun and her fingers gradually begin to respond. Exercising her hands she clenches and then slowly opens, clenches and reopens. Then like a conductor of a symphony or like a baseball pitcher warming up before the big game, she gracefully moves the yarn to and fro with her needle and feeble fingers. Following a pattern, she makes the most beautiful blanket for one her customers. Pausing for a second, she reaches for a ball of yarn. She then takes in a big breath of air and whispers to herself, "Oh, how time has flown! My Lord, I've gone through a lot."

"All I can do is think, yes think about the day," she whispers to herself. "Eat breakfast, take my pills, wear my darn hearing aide, which I hate and wait for Mike to pick me up on the TLC bus."

TLC stands for "tender loving care." It is a senior day center, where the terminally ill elderly meet and socialize and at the same time get medical attention. She has good times and tough times at the center. Her biggest obstacle was communicating with the other clients. On occasion she wears a hearing aide, but most of the time she'd forget or just didn't want to wear it. She loves to take samples of her crochet work to the center and many of the ladies would want one or two items. She takes orders. Most of the ladies would promise to pay. So when she'd come home she'd tell Louise, her daughter, that so many ladies ordered this and that. Because Maria had a mild case of dementia, Louise would have a hard time believing her and now that's where the conflict starts.

During this trying time the Crochet Lady has steroid medication injected into her ears. If any of you have experienced prednisone, you feel no pain and your energy level is high. She is like speedy Gonzalez, back and forth in her walker, breaking the speed of light. She can't sleep so she keeps everyone up, banging pans in the kitchen shouting out loudly. When she spoke to you she was two inches from your face. Surprisingly, for the first time I heard her say a curse word or two.

There are times it seemed like yesterday when she would patiently wait for her son to pick her up to go eat breakfast. How she looked forward to that special moment where they would cruise the city streets looking for that special place to eat and share the stories of the past. It usually was a greasy spoon, not the healthiest and cleanest of places. But it was all about the moment, those very last precious moments… It wasn't the best food but the best time to spend with her son, who'd been fighting cancer for so many years… Now he's gone and in a memory she treasures forever.

No mother should have to find her son in a nonresponsive state, but she did. Prior to his death she found him in his apartment with no movement. She called 911. He died weeks later at St. Luke's Presbyterian Hospital where the family had to stop all medication and machine assistance. This was a horrific experience for the Crochet Lady.

Once when she lived with us, she thought that her daughter Louise was dying. Early that morning she called out Louise's name and because Louise was in a deep sleep she wouldn't answer immediately. The Crochet Lady concluded that Louise was dead, so she called 911. The paramedics came, along with the fire department. It was quite the scene; the entire neighborhood was outside wondering what was happening.

After the fire department and paramedics left she went back to her normal self, crocheting as if nothing happened. She went to and fro, in and out where the pattern slowly became an illustrious mosaic of beauty. She pulled the work apart not meeting her expectations and started over. She then remembered that it was time to start a new colored line, so she slowly stood up and glided to her bedroom to get another color of yarn.

"The bus must be late, did I eat breakfast? I can't remember if I took all my medicine. Well, Louise will tell me. Should I tell Louise I fell last night? Maybe if I wear long sleeves she won't be able to see this big bruise. I just don't want my kids to worry…"

With the bus arriving shortly, so deep in thought she thinks about her attire for the day. "Do I need a sweater or big coat?"

In a flash she remembers the box of tissues to wipe her constantly running nose. At that point it's back to her position in the chair moving the yarn in and out. She then and there looks down and sees the toy train her great-grandson left on the floor. The toy train reminds her of a story a man told her when he sent away from home at age eleven and how he made his way from El Guique, New Mexico to Chama, New Mexico. She envisioned him running alongside the train, jumping up and grabbing the door's handle and dragging his feet on the ground until he pulled himself into the safety of the boxcar. He's my hero.

"Mom, Mike is here!" Shouts out Louise, "Don't forget to take your walker. I'll call Dr. Davis and ask him about your tests. Maybe less Tramadol so you won't be so dizzy."

The Crochet Lady responds with a quiet, "yes."

"Mom you didn't hear a word I said, because again, you're not wearing your hearing aid!" scolds a frustrated Louise.

Louise then races to the back room to search for the hearing device.

Finally, after ten minutes, Louise finds it, all mangled and twisted.

The Crochet Lady then walks slowly towards the front door taking her time gathering her crochet orders for the ladies at the senior center. Her loaded down walker was quite a sight to see. "Morning, Maria," greets Mike her bus driver.

Mike just laughs to himself and can't believe how much Maria has loaded up her walker. This is where she would disagree with Louise and because the Crochet Lady is the mom and Louise the daughter, Louise was to show ultimate respect even if it endangers the Crochet Lady's life. Keep in mind she was heavily medicated.

I first met the Crochet Lady back in 1972. I was a second year student at the University of Northern Colorado, and a beautiful girl caught my eye. I was invited by her daughter for dinner at her parents' home. Louise, her daughter and I had been dating for some time, and it was now time to meet my future wife's parents. The Crochet Lady and her husband must have been in their 50's and were mostly Spanish speaking. They were very cordial and hospitable. But what was so interesting about that meal that night was that the Crochet Lady prepared the most delicious stuffed pork chops with the best side of fried potatoes I've ever eaten. Later that evening, Louise told me that her mom had never made stuffed pork chops before. Wow! Was she trying to impress me?

Now the Crochet Lady is ninety years old, and there was no way that she was going to live independently. Losing her independence all started when the state took away her driving privileges. She was living in an apartment in Commerce City, Colorado, and was very used to doing what she loved best, whatever she wanted to do. But as age caught up to her, everything started to go-eyesight, hearing and mobility.

Albert Quintana

She was so proud of it so she framed it

CHAPTER TWO

L'Ojito, Colorado 1889

...Paradise...

The Crochet Lady gets on the bus and shares with the other clients that her story is a *corrido*, a song that tells a story of oppression, hate, pain, joy, love, adventure, and a lot of hard work. It all began in a small hamlet in southern Colorado called "L'Ojito," or known today as Jacobs Hill, located in south central Colorado. It is nestled at the base of Jacobs Hill; twelve miles west of Capulin, Colorado. L'Ojito is forest filled with the most beautiful pine, aspen and blue spruce trees to the west and desert like scene to the east. Also most apparent were ojitos, springs of artesian water and most predominately a stream that runs through its beauty.

Some people believed that it was once a hole in the wall, a place of hiding. What an ideal place to take refuge! One story from the past tells of a French miner who was passing through the San Luis Valley through Capulin on his way to Conejos County to set up a business project. The man came across a mother lode of gold and flaunted his riches in Capulin. One young man saw an opportunity to steal the gold from the Frenchman's possession. The young man got the Frenchman drunk and invited him to stay overnight at his cabin. In the middle of the night the young man got into the man's satchel and took the ore. The next morning the Frenchman woke up to an empty satchel bag. Immediately, the Frenchman questioned the young man. They tussled and got into a vicious fight. The Frenchman was much bigger man in stature so the young man equalized the fight by using an ax. Needless to say, it ended in the death of the Frenchman by a brutal ax blow to the head. Two weeks later, a group of French miners came in search of their partner. The people of Capulin were not too fond of the young man who killed the Frenchman. Slowly the people of Capulin noticed that the young man was buying land and flaunting the goods he had from his new wealth. On several occasions he would even brag about his feat in slewing the European from across the big pond. So the more he bragged about killing the Frenchman, word eventually got out to the sheriff. Later they found out that the sheriff was friends with the deceased miner. Hence, from Capulin this young man fled to L'Ojito in hope he wouldn't be caught. Six months later the man was caught, tried and hung. Behind the cabin of the young man's home, the sheriff found the Frenchman's body partially buried. The young man was found wearing the Frenchman's clothes and boots. The young man had a size 5 foot but was wearing the dead man's size 9 boots. The boots were filled with newspaper to make up the difference.

"He was one of my ancestors," explained the Crochet lady. "He was the last man to be legally hung in Colorado. As the historians put it, at the gallows prior to his death he asked for the forgiveness of all present. His defense in court was self-defense but it did not stand up in court."

"Wow Maria! You are quite famous," said an astonished Alvie, the Crochet Lady's best friend at the center.

"I don't know if it's being famous, probably more like not so famous, infamous!" countered The Crochet Lady with a warm laugh.

"Yes L'Ojito was quite the place. It was a paradise in the eyes of a mountaineer. We were very, very, poor, salt of the earth poor. Just imagine we survived the Post-Depression years as Lumberjacks, migrant workers and housekeepers. We worked where there was opportunity, where ever Petro would find work. Yes you wouldn't believe it. My three half-sisters and I were lumber jacks. Many called us 'Tom boyish' and others called us 'Lumber Jacquelines.' We worked very hard cutting down trees marked by the forest department. We could only cut down those trees. Most of the trees that were cut down were infested with beetle kill," explained The Crochet Lady. "Because of the Beetle Kill, we made our living. With two-man saws and occasionally when the power saws were working we'd cut down the trees. Physically, we'd get down as low as we could, to get the most of the tree. Nothing was wasted. After cutting, we'd load the huge logs onto the truck and sometimes into a trailer. When the truck was not operating the horses would bring the logs down from the mountain top on sleds and then we'd unload at the saw mill. It was very physical painful work! As a result of that physical work I now have two artificial knee joints.

The ladies at the center were amazed at the Crochet Lady's story. They all gathered around her and wanted to hear more.

"Tell us more Maria, you are quite a unique person with such an interesting history, tell us more," shouted the ladies in unison.

But she could not hear very well and sometimes answers incorrectly. "Well maybe tomorrow, I'm really getting tired and I'm losing my thought." said an exhausted Crochet Lady.

So the Crochet Lady sped off to her favorite lounging chair and within seconds she sleeps. This was pretty much the routine of all the ladies and men at TLC. Twenty minutes is the normal length of daily naps.

When she awoke she reaches into her red crochet bag for her tools of the trade. Then like a condor ready to take off for flight, she focuses on the task at hand. She selects her yarn color from the bag, takes out the needle, stretches her fingers, and with such fluidity, works her talent. What a beautiful sight to see. It was like Leonardo da Vinci painting the Mona Lisa, or Michelangelo painting the ceiling of the Sistine Chapel.

Right after lunch all of the ladies gathered around the Crochet Lady and wanted to hear her story.

"Tell us more Maria, what about your childhood," shouted out an excited Mildred.

So the story continues...

The Crochet Lady

The Year was 1922. A young Maria at 5 months of age embraces her mother at the burial site of Juan Pablo her father. He fought a lengthy battle with diphtheria. Diphtheria was a terrible and painful way to die for her dad. You see, the throat swells up until you can't breathe anymore.

Surrounded at the burial site also were her eldest brother Regino, Uncle Petronino Lucero, Grandpa Cayetano Lucero, Aunt Francis Lucero, Uncle Maria Gomez, yes Uncle Maria because his parents named him Maria after the Blessed Virgin Mary, Uncle Jose Lucero, Uncle Felix Ortiz half-brother of Juan Pablo Lucero, (the direct off spring of Ortiz who was the last man legally tried and hung in Colorado.) Also amongst those in attendance were his wife Constancia Lucero Solano, my mother, Grandma Susana and other relatives and friends from the San Luis Valley.

It was a very sad day for everyone. Juan Pablo had just recently bought the future land where his wife and children would live, better known as L' Ojito. Constancia and Juan Pablo had planned on raising the children and developing the small ranch into a larger one, with more horses, cattle, and lots of chickens.

Petronino was very sad at the death of his half-brother who he looked up to. He knew that his brother's wife Constancia would have it rough raising two children and tending the ranch that was left back for Regino and the Crochet Lady. But the Crochet lady was too young to understand what death was about and only knew that her loving father was no longer present in her life.

"What will we do without father? Your father had plans to rebuild the ranch. His dream was to leave it to Regino and Maria. You are so young as the work is so hard for us," cried out a troubled Constancia.

Times were very hard for Constancia. Juan Pablo had left her with some cattle for meat and some chickens for eggs. But as there was no money coming in, Constancia would have to sell one cow at a time. Yes indeed, it was a very sad moment. Constancia always believed that God answered prayers. Her prayer was to find someone who would help her raise her two children and develop the ranch.

Constancia had a great faith in the Lord Jesus Christ and in La Virgen de Guadalupe. She would regularly be seen with her hands wrapped with beads and prayed the Rosary on a daily basis. With her persistence and many "Hail Marys" and "Our Fathers," would her prayer request be answered?

CHAPTER THREE

A Dilemma

"... Sometimes in life one must choose between alternatives that are less desirable, but the outcome is worthy..."

Months later after the death Juan Pablo, Constancia suffered great depression. Her parents kept her and the children with them and just hoped and prayed that Constancia would find a husband to support her soon. It was so sad to see their daughter in law in such grief. Cayetano was not the richest man in the Capulin area, but did have some money set aside to help develop the ranch at L'Ojito. So Cayetano's request was to save the ranch at L'Ojito by having his youngest son Petro marry Constancia and take over its ownership.

Petronino was still very young in fact he was only four years older than Constantia's eldest son Regenio. Maria and Regenio only knew Petronino as their Tio. (*uncle*)

"Petro my son, your brother's wife is in turmoil. I am asking you to marry Constancia. She still has many child bearing years left to give you children and probably a son to carry on your name," said a sympathetic Cayetano.

"But papa, I want to marry Gertrudis Sandoval from Antonito, I love her!" cried Petro.

"The time has come and we as a family must support one another, replied his father.

Trust in me, my son and all will work out. Regenio your nephew will be a man soon and he will be able to help support the family."

"But father, this is a lot of pressure on me. What will I tell Gertrudis and her parents?" questioned Petro. "Please give me some time to think this one over."

Two weeks had passed and Petro succumbed to his father's wishes. So hesitantly, at a very young age of 19, Petro took on the responsibility of marrying Constancia and the raising of his nephew and niece Regenio and Maria Ida.

Maria Ida and Regino never called Petro, father, but tio. They only knew him as their uncle. This never became a problem for Petro, because he never referred to them as his children.

I'm sure that Petro had gotten permission from the Church to marry his sister in law who was ten years older than he. Kind of different, but it was common practice in history. When a brother would marry the widow of their brother, he would be responsible for the support of the widow and children. But in this case the motivating factor was Petro would have ownership of the estate at L'Ojito.

Gertrudis Sandoval and her parents were so upset with the decision Cayetano made.

Back at the senior center...

"My mother married my Uncle Petro," said a hesitant Crochet Lady. With a very attentive crowd the clients couldn't get enough of the story she was telling.

"Tell us more! Tell us more!"

So there I stood as a little girl somewhat confused about how my Tio Petronino, was now my Step Father. Was it possible that I was the only one who thought that it was strange? The other students at the Gomez school would sometimes make fun of me and say that it was wrong that that my step father was my uncle. That's impossible; you're just a huerfana (orphan) with no daddy. I tried to defend myself but no one would support me on the playground. The shouts got louder, "Huerfana, Huerfana!"

My first day in school was a disaster. All of the kids called me names and the crazy thing about the whole thing was all of the students were my primos. (cousins) I cried all day long and the teacher tried to find out why I was crying but I wouldn't tell her. I couldn't wait to get home to tell my mom.

Mom always said the right things to make me feel better. I told her the story and told she told me to just ignore them. To ignore them would be a hard thing to do. But I do remember Tio and Mom going to each of the vecinos (neighbor's) houses to give them some shit. Well, that's the word Tio used to describe his disagreement with the neighbors. The next day at school was good. All of the students were so nice to me. The rest of the year went just fine.

CHAPTER FOUR

Mount Blanca 1928

Legends are stories passed down from generation to generation.

1927 was a remarkable year for us in L'Ojito. I now have a baby sister named Catherine, named after my Great Grandma Catherine Solano. Catherine's birth was a great celebration. The sad thing is that my brother Regino no longer wanted to be part of the family because he didn't get along with Tio Petro, so he ran away. My mother was so sad but she understood that he would have to be on his own at the age of 16. But on the other hand Catherine brought joy to us all. She was such a cute baby and I loved her so much. I loved to hold her in my arms and squeeze her and give her bunches of kisses. Petro was so excited about his new baby girl and spoiled her to no end.

Toward the end of the school year right before summer vacation a very interesting man with a headdress made of feathers came to visit the classroom. Mrs. Garcia, our very nice teacher said that he was a Navajo Chieftain. His name was Doko, translated as the "shelled mountain." He was so interesting and entertaining. He told us about the legends of the four surrounding sacred mountains, Mount Blanca, Mount Taylor, Mount Humphrey, and Mount Hesperus. These were all of major mountain peaks that were near us. I loved the stories about the peaks but what really stood out to me was Mount Blanca the White Shell Mountain. I knew about Mount

Blanca because it was always the whitest of all the mountains. Chief Doko understood that the great creator said that Blanca represents "thought." He continues by saying that thought or reason comes first in everything you do. So the creator carefully formed the mountain by giving it spirit and from the spirit grew its appearance. So it took shape with a dome of white and beautiful plants and flowers.

He said that the White Mountain represents life and everyone must have thought to make decisions. Our decisions develop our lives. He continued and told a story about a young warrior who lived a far distance away and was in turmoil, his Father no longer wanted him because his appearance was different from all the others in his tribe. His father accused his mother of not being faithful to him and sharing her bed with another when she went to the rendezvous to get supplies from the Frenchmen. When the young warrior was born he was light skinned and blue eyed with reddish brown hair. The Chief named the young warrior, *Tsisnaasjini* (White shell mountain) or as we know, Mount Blanca. All the other Indian braves made fun of the Chief and told him that he shared his bed with others. Slowly the great chief was losing respect from the others, so he killed his wife and beat White Mountain severely and banishing him the tribe. This was the ultimate punishment, to be forced out. Consequently with a lot of thought, Mount Blanca traveled north to find his permanent home. Therefore, that night in a dream the Great Creator told him to travel north where he would find a flock of sheep drinking from a stream. It would be surrounded by beautiful trees. Amid the trees you'll find a beautiful maiden who will love you with all the love you need to mend the scars from your past. She will be lovely and full of thought. She will give you beautiful children and grandchildren.

The days had passed and the trip north was grueling. Hunger was something fierce and the surrounding tribes were at times friendly and others were ready to kill. But Mount Blanca was brave and knew that there was a treasure at the end of his journey. So one morning, just as the Great Creator promised in the dream, White Mountain walked through the forest and out of the woods. He found one sheep, then two sheep and down in the valley was a whole flock of sheep. He walked slowly through the grazing sheep and stumbled over a rock. When he lifted himself from the ground, there she was, the most beautiful woman he'd ever seen in his life. She had long black silky hair that glistened in the sunlight. Her complexion was flawless, so smooth and creamy. Her beauty was beyond belief. There she was amongst the pine at the side of a stream. She was a Ute maiden and she called herself, White Butterfly. She too like Mount Blanca was light skinned and was banished from the tribe because she did not want to marry a brave her father had chosen for her. White Mountain was astonished and couldn't believe that the Great Creator had guided him to White Butterfly through a dream.

"Great Creator, my Creator why am I receiving such grace? Cried out, a very, very happy White Mountain. I am not deserving of such love."

Then the two journeyed northward to find their new home. White Butterfly's parents got word of her exile and wanted her back with the clan. Knowing this the two fled northward to find their new home. The journey seemed like years, but the further they drudged northward their love grew stronger. Quickly the years passed. Slowly they grew old. White Mountain was losing his hair and White Butterfly's bones started to ache with pain. The days grew shorter and the nights longer and then the blizzards came. They walked and walked for many miles. The snow became deeper and deeper. Gradually, the two could feel the freeze. White Butterfly shivered and fell numerous times. Finally, they could not get up. White Mountain stood over his Princess and then bent over her like a shield protecting her from the whiteout. Time passed slowly, the two were conjoined frozen, solid as one. So, to this day when you look up northward think about the Legend of White Mountain and White Butterfly.

The student's eyes were glued to the front of the classroom and were so intrigued by this beautiful legend. At the back of the room was Ms. Garcia sniffling and trying to hold back the tears, "Oh Chief, that was so beautiful."

"Chief Doko, when can we have you back to share more legends?" asked Mrs. Garcia.

"Anytime," responded the Chief.

"We sure would like you back Chief Doko," responded Mrs. Garcia

"My office is located in Alamosa at the College. So I could visit at any time," said Doko.

"Wow, what a beautiful legend, I was so touched by this story and wished that one day; I too would find my White Mountain," replied the Crochet Lady.

When the Crochet Lady got home after school she couldn't wait to tell the story of White Mountain and White butterfly.

"Tell us the story *mi hita* (daughter), I am excited to hear what you learned in school today," replied her mom.

"Tio, do you want to hear the story I learned today?" she asked.

"No not really, he answered, I have too much work to do around here since Regino ran away. Now it's just too much for me to handle so I'll have to go into Monte Vista and find a job until planting season comes around and the entire family will have to follow the crops. We will pick peas, pickles and in late summer potatoes. The pay isn't that good but with the three of us we should be able to make it through next winter."

Come on Tio, it's a good story about the White Butterfly and Mount Blanca," explained Maria Ida.

"Oh, okay now quickly, because I'm meeting your Grandpa Cayetano in a few minutes."

So the Crochet Lady at age six told this beautiful story and her Mom thought that it was so heartfelt. All you could see was a stream of tears rolling down her cheeks. Petro just was impressed but set back and in his own macho style replied, "Pura mentieras" (Pure lies)

"But its true Tio, a great story which will always be in my heart,"

At that point the Crochet Lady looked over to her little sister Catherine and said, "I can't wait until you are old enough to tell you the story about the White Mountain and the White Butterfly.

"Wow, that's a long time ago and now I'm 90 years old and can barely walk and hear. Damn it! I hate that hearing aid and more so the walker. I get so easily frustrated when I can't do what I use to do. Slowly I'm losing my crocheting skills and losing track of where I am in my crocheting patterns. I just noticed that Albert, my son in law brought in a wheel chair. Now, I will not have anything to do with that damn chair! As soon as I start riding that chair I'll become lazy and slowly my body will become weak and I will die. I do not want anyone to feel sorry for me, No One!"

The next day at the center the Crochet Lady received news that another client at the center had died. It was a sad day and everyone was so melancholy. Not too many wanted to hear the Crochet Lady's story, except one man who occasionally would ask questions about her family.

"Maria, please continue this beautiful story, please," he pleaded.

CHAPTER FIVE

The Lion's Gate

128 miles to the south in Tierra Amarilla, New Mexico Court House, stood a very shy and extremely sadden young boy. He was wearing some much worn overalls; beat up shoes and a faded red flannel shirt. He was a handsome young man. His most striking features were his reddish brown hair, bluish green eyes and most distinctly, a broken left arm that atrophied with no ability to move. The arm was stiff and just dangled. It was never set in a cast to be fully healed. He was brutally beaten by his father for not doing his chores. Obviously, this had happen many times before, because now they were in court.

At the age of eleven he tried to be strong, but was scared to say the least.

His cry was so deep with pain.

"Why does it have to be this way?"

He sat in the chair with his hands covering his eyes. It was time for the judge to make a decision. Should the boy be put in an orphanage or should his father Enrique, be sent to prison for over disciplining him with a shovel? The boy's name was Pantalion.

"Pantalion," the judge said. "I have gone over your case thoroughly and I have decided that you will live in the state orphanage in Santa Fe for three years where you will get an education and three meals a day. If someone wants to adopt you, you could be adopted. Then after the three years, if no one wants to adopt you can go back to your family in El Guique.

Pantalion's mother Roseana screamed out with such pain and pleaded out, "No judge, *es mi hijo, mi nino*, No! No! Please judge no! He's my baby!"

All of Pantalion's relatives were there in tears. Crying, everyone was crying, aunts, uncles, grandparents, and all of his five younger siblings. The court bailiff handed out paper tissue to everyone, upon waiting for the judge's verdict. It was quite a sad moment to say the least.

Finally, after 30 minutes the judge came into the courtroom. It was now time for the deliberation.

"Enrique, you will be on probation and will report to the sheriff in Espanola monthly until your son is released from the orphanage in Santa Fe. Mrs. Sisneros you will be able to visit Pantalion as much as you want." "Mr Sisneros, If you didn't have such a big family to support I'd send to you to Canyon City prison in Colorado for 25 years. I expect you to treat your children with love and care. Ultimately, I do expect a positive report from the Sheriff in Espanola."

During the entire deliberation, Enrique fidgeted and sweated profusely. When it was time to leave, Enrique was asked to go back to the judge's quarters where he was spoken to at length on how to civilly discipline his children. In the courtroom the other family members gave their farewells. Roseana, his mother promised that she'd go visit him at the orphanage. Pantalion held his mother in a tight embrace and eventually broke away and went with the social worker to the bus.

The bus ride to Santa Fe was a long drive and all Pantalion could do was think about his mother and siblings. It was a good long cry as he held his hands to his face and sobbed out loudly. Mary Pendleton his social worker rode in the bus and sat next to him and tried to console him, but all he could do was sob uncontrollably.

"Pantalion, you will do just fine in Santa Fe at the orphanage. There will be other students your age that you can play with. There will be classes where you'll be able to learn how to read and write. You'll have your own bed and not to mention we have a great cook who makes the best tortillas, frijoles, posole and all your favorites and some American food like hamburgers and french fries. Have you ever tried hamburgers?" asked Mrs. Pendleton.

"No senora, I've never eaten hamburgers, but I've heard about how good they are." answered Pantalion.

"Well, Pantalion I've studied Latin in school and learned that your name has a very special meaning," pointed out Mrs. Pendleton.

"Mrs. Pendleton, what does it mean? My parents have never told me its meaning. All they told me was that I was named after my great uncle Pantalion Garcia"

"Your name means the 'Lions Gate' which means that you are very strong protector, she explained. This means that for the next three years you will be strong and intelligent. We at the orphanage will make you feel special and loved."

This assurance made Pantalion feel good inside. The caring he never ever received from his dad. There was hope for Pantalion. Could this be a blessing?

The ride to the orphanage was made a little more comforting after Mrs. Pendleton's kind words. But her kind words could never replace the love that he had for his mother, brothers and sisters. They were truly missed.

CHAPTER SIX

New Mexico State Orphanage

Back at the senior center...

Gradually, the other clients at the center gathered around the Crochet Lady. With close attention they listened to the story about Pantalion.

It was Friday afternoon when he arrived at the orphanage in Santa Fe. Gazing out the window of the bus all he could see was the biggest building he'd ever seen in his short life. The entire building was made of a red mud material called adobe brick and was probably five stories high. Behind the building was a huge football field with a track. He entered the building through the main entrance and was directed to the building's check in room. The checkout room is where the new clients would be given the essentials to a life at Santa Fe Orphanage.

"Young man, what is your name, *que es tu nombre*?" asked the check-in supervisor.

"Mi nombre es Pantalion Andres Sisneros," replied a shy, not too confident, but subdued Pantalion.

Pantalion was somewhat surprised to hear the checkout supervisor speak

Spanish so well.

"My name is Manuel and I speak Spanish because it's my first language like you Pantalion.

But, here we speak English, so we'll call you Andy, it will be easier for you to be acclimated into the system," he responded.

"So Andy it will be, Andy Sisneros. That has a good sound to it."

"Here you go Andy, two pairs of pants, two flannel shirts, and boots. Two pairs of underwear to be washed daily, a set of sheets, and a wool blanket. We also have to include all of your toiletries, toothbrush, comb, bar soap, and two rolls of toilet paper."

"You will be responsible for everything in your arms and every Friday is sheet day. So on Fridays you will place your dirty sheets at the foot of your bed. Do you understand? Friday is laundry day."

"Yes Manuel, thank you."

All of this was new to Pantalion, especially the toilet paper thing because he was accustomed to the Sears and Roebuck magazine sheets used at their outhouse.

From there Andy was directed to the dormitory where he would find his bed. The room was elongated with two rows of six beds. The beds were big enough for one person, not like at home where he had to share his bed with his two brothers. This was something new for Pantalion; it was a good feeling to have something of his very own. At that very moment the coordinator of the orphanage walks in and startles Pantalion.

"Good afternoon young man and who you might be?" he asks.

Mi nombre es Pantalion Andres Sisneros but the man said, "Only English."

"My name is Andy Sisneros," replied Pantalion.

"So we'll call you Andy, yes Andy, a good American name. My name is Mr.

Mc Farland and I'm your coordinator, kind of like a principal of a school," introduced Mr. Mc Farland.

Mr. Mc Farland was a big man, about 6'3" and weighed 250 lbs. He spoke very deep and loud. He had big hands and walked with a limp. It was probably from a war injury. You could tell by just the way he spoke that he was the man in charge. He demanded respect and got it.

"Andy, you're probably wondering why I walk like this, well it was the war, WWI. I was hit by some shrapnel from a grenade in the trenches in Germany defending our country," informed Mr. McFarland.

In amazement, he stared at the director for a few seconds. He wanted to ask what WWI was and what the heck a grenade was. But, instead he just listened and marveled at this man's intelligence.

Andy now was directed to the big room on the main floor where he met all of the orphans. There were children from ages of a year old to 16. There must have been around fifty children in the big room. From that point on they all called him Andy.

It was now time to go to bed. Here at the orphanage the toilets were inside not out. To Andy's amazement he couldn't believe that people used the toilet inside and not outside. There was actual running water where people could wash their faces and bodies. Back home he'd have to carry pails and pails of water from the river to drink, wash and bathe. It was such a hard task. But here everything was at our finger tips. At home we bathed in big tubs but here we took showers. This was going to be like paradise.

That night he slept in the most comfortable bed that he'd ever slept in. There were sheets and very soft blankets and very soft pillows too. In the bed next to Pantalion was Jesus Vigil, a boy about his age.

"Hello Andy, my name is Jesus Vigil. Everyone here calls me Jessie. I am an orphan and just waiting for someone to adopt me."

"Good to meet you my name is Pantalion, I mean Andy. Just call me Andy. They don't want me to speak Spanish."

"Don't worry Andy, when they are not paying attention we'll speak Spanish, the language of our people," said a proud Jessie.

"My parents died in a car accident when I was 3 years old. I've been here for 8 years and hope and pray that one day someone will adopt me. Every Saturday people come in and look us over for adoption. Sometimes it feels like we are in a pet store. Most of the younger children get selected over us older children, probably because they are cuter than us older children," explained Jesus. But, I think it's my ear, because when we crashed in the accident part of my ear was severed, cut off. Now that's why I wasn't adopted. I once went home with a young couple from Santa Rosa but they brought me back because that same month the lady said that she found out that she was pregnant. It was a sad thing for me because I thought I was going to have parents to love me," explained Jessie.

"Well Jessie, I don't want to be adopted. I have parents back in El Guique. I've been court ordered to come here because my dad hit me with a shovel and broke my arm. He hates me and treats me like trash, like I'm not his son," explained Pantalion.

Then there was this complete silence between the two. The discussion brought back bad memories for Pantalion. He went into a deep depressive state. All he could do was think about his mother's crying in the courtroom. Jessie was very respectful and knew that Pantalion needed time to heal his broken heart. Jessie could see that Pantalion had covered his head with the blanket. He didn't want Jessie to know that he was crying.

"Don't worry Pantalion, I mean Andy, I will be your best friend and I'll show you the ins and outs of this place," assured an empathetic Jessie.

At that moment from the very back of the room in bed 6, came a loud voice.

"*Callete! Callete*! Shut up! Shut up! You little pussies need to get your asses kicked. I need to get my rest, so in the morning you'll feel the wrath of my boot."

Jessie then whispered, "That's Billy Caldwell the orphanage bully who loves to fight. I'll teach you how to survive his bullying."

"Bullying, Bullying, *que es* bullying Jessie?" asked a very confused Andy.

"Bullying is where one guy picks fights with others for no reason. Billy will definitely try to fight you to see how tough you are. He tries to dominate everyone. Don't lower yourself to his game."

"Okay my new friend. I think that we are going to be great friends,"

The next morning the sound of a bell filled the room. It was time to get up. A voice shouted out, "get up, breakfast in 15 minutes."

Pantalion was somewhat confused. When he was home at the ranch he'd just get up, put a couple of logs on the fire, go to the outhouse and then feed the cows and the chickens. Much later he'd have breakfast. Here, the twelve boys lined up at the rest room and showers and off to the eating hall they assembled.

Now the orphans were required to say grace before the meal usually led by Sister Ann Castillo, who was a volunteer from the Catholic Church down the street. She was so nice and pleasant and also asked the Lord during prayer petitions to heal Andy's broken heart.

Jessie was a great mentor for Andy. Andy just did everything Jessie told him to do. Then after the breakfast they assembled in the courtyard for Physical Education. Every morning they were required to run a mile around the track. Andy had a hard time running but after a week he was able to run the mile easily in seven minutes.

The rest of the day was spent in the classroom learning how to read and write. Andy's experience in the classroom was minimal, for he only attended school to the first grade and that was sparingly. This certainly would be a challenge.

The next day after breakfast Andy was introduced to the orphanage bully Billy Caldwell (the oldest orphan in the facility.) It was a punch to the kidneys and flick to the ear and a how do you do, red headed Mexican?

"Damn, I didn't know they made red headed Mexicans, you must be a, what my dad use to call, a 'Coyote', taunted Billy Caldwell. You have to be a half breed."

Andy turned around and saw the biggest, meanest, ugliest kid he'd ever seen in his life. Billy must have been 6 foot tall, 250 pounds of fat and the blondest hair he had ever seen.

Perplexed, Pantalion just froze and couldn't move from the floor. He didn't know whether to run or cry. But something inside him told him not to move.

Jessie stepped up and said," Billy leave him alone he's new and you know what happened to you when you picked on the last kid here."

Billy had quite the dysfunctional life. He was in and out of several relative's homes but they could not raise him because he was just a very angry kid. He never knew his mother. His dad was in prison for manslaughter and robbery. His relatives didn't want him around. They all said that he was a nuisance.

Billy loved his title of being the 'school house bully.' He was a master initiator. He hit and teased the others when no one of authority was watching. He had the record for the most swats to the buttocks for his misbehavior. Earlier in the year he hurt a kid badly breaking his arm. He then was put on probation. Billy ended up being arrested. The sheriff warned him that the next time he got into big trouble like a fight or stealing he would be sent to the state prison for troubled teens. He just loved to fight and make others cry especially the girls. He just loved to pull their hair and say nasty things about them. The girls were never allowed to be left alone with him in the same room. But this didn't stop Billy from harassing the others.

"I just want you to know that I'm the main chingon (boss) here and you better do what I say or else there is going to be some 'ass whoopin,'" challenged Billy.

Jessie picked Andy up off the floor and assured Andy that one day Billy would meet his demise. Andy thought, what's wrong with this guy I don't even know him and he's punching at me and making me very angry. I'll just have to stay clear of him.

After the ordeal the boys went to the wood shop where the students would learn how to make furniture out of wood. This was going to be a great place to learn how to build, Andy thought.

"I would love to build one of those cedar chests that smell so good for my Mom when I get out in three years."

The woodshop was set up in two different rooms, the front room was the classroom where they would study and take notes on all of the demonstrations. The back room was the wood shop where there were power table saws and drills. There were also many kinds of tools which included hand saws, screw drivers, hammers, measuring tapes, planes to shave wood smoothly just to name a few. To make the projects look good there were paints, varnishes, shellacs and oils which really brighten up the woodwork.

Pantalion was in his element. He could remember seeing his grandfather build chests out of cedar. This was his place of refuge where he could get away from all the bad memories from the past.

The second day was a pretty fun day. I'd say the second day because the woodwork class was taught every other day and exchange with the music class. When we walked into the class the teacher would take roll. It was funny to hear the way the he would pronounce the Spanish surnames. The students would have to correct him repeatedly. The teacher always butchered his name Sisneros. Every once in a while he'd call him Pantalion and then correct himself and call him Andy.

"Sorry about that Andy, I need to change and correct the roll call list and pencil in Andy," said Mr. Villani.

Hearing this, Billy felt that it was time to tease Pantalion. So when class ended, Billy waited for Pantalion outside of the classroom. Then the harassment started.

"Pantalion, Pantalion, hmm, Panties, Panties, that's a great nick-name for you Aka Andy," teased Billy incessantly.

The anger within Pantalion was at boiling point. His face was beet red and his forehead dripped with sweat profusely. He retaliated by yelling out,

"Cabron, Puto!" (Goat whore!)

Billy squared off and stepped up and held up his fat pudgy fists and said with snot running from his nose, "okay now it's on, you little red headed beaner!"

At that instant Jessie jumped up to calm down Andy and said, "Wait my friend he'll get his soon."

Mr. Villani noticed the raucous at the door shouted loudly, "Knock it off, damn it, and knock it off!"

The boys walked down the hallway and then Billy put his face into Pantalion's face. He then grabbed Pantalion's disfigured arm and commenced twisting it. In anguish Pantalion pulled away and backed off some distance. Billy expected Pantalion to cry, but he knew better to not cry out loud, because to cry out loud would show weakness. Even when his Dad would punish him, he'd stand strong to show him that he was a man. That's probably what got Enrique upset the most.

His dad Enrique Sisneros was a pretty tough little guy in El Guique, no one messed with him. Pantalion could remember when one day his family went to a party in Espanola. Some guy from Texas with a big mouth was talking dirty to his Mother. His dad was somewhat diminutive in size. He stood 5'3" and about 140 pounds. He got so pissed he hit the man with a bottle of whisky over his head and commenced striking him with 500 punches or so. He was one bad ass, but he didn't love Pantalion for some reason. He didn't think he was his son. Sometimes Pantalion would be the punching bag when Enrique was upset about anything, anything.

After the scuffle in the hallway the boys went on their own way to the next class. It was a hard time for Pantalion to survive at the orphanage. Billy was constantly teasing him. If he saw Billy he'd go the other way. When it was time to go to the cafeteria he'd sit in a position where he knew Billy was at all times. It was miserable. The taunting would have to stop.

Back at the Senior center...

"Ok clients, it's time to get on the bus to go home," informed Juanita.

All of the clients looked over to the Crochet Lady and said, "please continue the story tomorrow."

The clients all gathered in single file and entered each assigned bus. Every once in a while someone would forget the bus they were supposed to get on. The bus drivers were so patient with the clients it was amazing. Mike is the best; he is very quiet and never talks to anyone but is the most caring when it came to compassion for the clients. It is always safety first. When it came time to bring the Crochet Lady home he made sure that she was taken care of. She'd get off the bus slowly with her walker and he'd follow her to the front door to make sure that she would not fall. She'd get the keys out of her purse and he made sure that she'd enter the house safe and sound. Then he'd sit her down into the front room recliner. Her daily routine was always the same. She'd get up, freshen up, take her walker to the kitchen, eat breakfast, take her pills, push her walker to the front room recliner and wait for Mike to pick her up. Then there was the long drive to the center and one at a time they'd pick up all of the clients. Their time at the center is 4 hours and then off they'd go home. Her home life was very much the same. She really liked my cooking but recently the food didn't taste good like it used to. After the dinner meal she'd get her crochet materials out and crochet a green doily. For some reason she'd forgotten how the pattern was supposed to go. When she finally remembered the pattern she was totally exhausted mentally. On many occasions she'd drop the needle and then she'd bend her neck over and would say that she had great pain in her neck and head. She'd then take her medicine for pain and then off to the bed. Hence the routine would start all over again the next day.

The following day the clients gathered around to hear her continue her story. She knew what they wanted, more stories about Pantalion. But she politely told them that it was a hard day for her; that her hearing was almost all gone and she wasn't feeling well.

A little old man who wanted to hear the story to continue, said, "Maria, please carry on with the story. It's so very interesting."

All she could hear was rumbling sounds in the background. She looked at the little old man mouth out, "Maria, continue the story please it's so interesting."

So without delay, she'd say, "Okay where was I?"

The rest of the clients gathered around and were elated that she would continue the story.

"The orphanage, Maria, the orphanage, Pantalion is at the orphanage," shouted Alvie.

He loved it at the orphanage; he was eating well and gained 10 pounds in the first month. Everything was great except for two things, Billy pestering him on a daily basis and the fact that no one from El Guique had come to visit. Nevertheless, he did get letters from his grandmother and mother. They always said that they'd visit but something always came up. Usually it was that they didn't have enough money for gas. Even so, they sent him their love. After reading the letters he'd get into his own little world and that's when everyone would have to give him his space.

CHAPTER SEVEN

100 days

We should never out live our children.

The Crochet Lady stopped telling the story about Pantalion and then there was complete silence for the longest time. At that point, with a great sigh she cries out loud, "My son, why, oh why did he have to die such a horrible death?"

"What's wrong Maria? Asked the seniors, you were talking about your friend and now you're talking about your son?"

"I'm going to have to stop for a while; I miss my son Alonzo so much. He died from a terrible death fighting cancer. No one should outlive their children."

Reaching for her box of tissues she delicately wipes her nose, pauses for a second, clears her throat, grabs more tissues until the box empties out and continues the story. I don't know what it was about tissues, she would get agitated and then all at one time she'd pull tissues out of the box until the box emptied.

Before I moved into my daughter's house, I lived at a senior living complex in Commerce City. My son Alonzo also lived in the same facility. I always wanted to get a two bedroom apartment so we could live together

The Crochet Lady

and share the expenses. He had Hodgkin's and Non-Hodgkin's cancer and was on a lot of medication. He had a bone marrow transplant that wouldn't graft. The procedure was 100 days of close care making sure that he was taking the right medicine, a sanitary environment and the right foods. Louise my daughter, Dan and Diego his sons took shifts making sure that he would follow all protocols. It was a very arduous and time demanding procedure. Everyone sacrificed their time for this procedure.

After the 100 days, he and I would meet at the lobby downstairs of the complex and go out for the breakfast. It was a very special time for us because we would share our ailments and stories of the past and have breakfast at every greasy spoon in Commerce City. It was a lovely time. He was in pain from all the chemo and I was in pain from my body tearing down. I loved him so much and would help him as much as I could. He was so talented. He could build anything. All of my children are very talented. Andy is a very talented roofer and Louise was a teacher who could teach anything. I never tell them that I'm proud of them, but I'm so proud and I thank the Lord for them.

"Oh, where was I?"

She then paused and grabbed her forehead with both hands and said, "*Que duele mi cabeza,*" (oh how my head hurts.)

Oh, now I remember where I was. Slowly after the 100 days Alonzo would forget to go to his doctor's appointments. His memory was deteriorating and he was forgetting to take his medications. His body was taking on many of the ailments of his donor. He then had acquired Diabetes, Neuropathy and Shingles. He was in great agony. His health was worsening and sporadically he'd have to go to the hospital for weeks at a time. After his last visit he looked great and healthy and then a month later, he was back to square one. It was a hard time for us all. Then out of the blue, my husband's sister Maria died in Espanola. Alonzo adored his aunt in Espanola and wanted Louise to go with him to the funeral. She couldn't, she had to stay back to care for me because I was very "sickly." When Alonzo came over to ask her to go to the funeral, he'd had gained about 30 lbs. He looked like a totally different person. He was always very thin. He had gotten balder from the chemo treatments and lost his teeth. We couldn't believe our eyes. He did go to Espanola alone to visit the funeral and cousins. Unfortunately, it was about a week later that I found him in his apartment, non-responsive. It was a very hard time for me and in a matter of days he died at St. Luke's Hospital/Presbyterian in Denver, Colorado.

Then she took a long pause, cried out loud and couldn't continue the story. All of the clients were sympathetic and some even cried along with the Crochet Lady.

Actually, she stops sharing her corrido for about a week. A week later she concluded the story about her son Alonzo.

We were all there for Alonzo's farewell. He was on a bed at the hospital hooked up to a machine that was breathing for him. I truly believed that he had already passed on to the next life. His family made the decision to take him off life support. Later he took his last breath. It was such a sad scene to see everyone in such turmoil.

My son's services were like a blur. I can't recall anything, just the music and the relatives who came to console me. Oh, I remember that Danny my grandson gave a great eulogy. A week later I was totally moved into my daughter's home where I live presently in the master bedroom which had everything on one level. That was good because I wouldn't be able to climb the stairs to the upstairs bedroom.

My hearing slowly worsened. Louise took me to the ear specialist where the doctor gave me steroid treatments. I did not like the results of the treatments because it gave me too much energy. I was constantly moving back and forth. On occasion I would want to leave their home because I missed my independence and freedom. I'd say that during this time I would get flash backs of seeing my son Alonzo not responding and lying in bed lifeless. But then again, in reality at my daughter's home I believed that my daughter was not sleeping but dying too. I called out to her name to wake her up and if she wouldn't answer immediately, I thought that she had died also. So, I called 911 and the police and ambulance came. It was a terrible feeling inside of me. If I had a disagreement with Louise I usually would just get on my walker and head down the street. I was uncontrollable. Louise would have to call the police and the police would have to calm me down. I frequently would tell the police all the things that Louise would not allow me to do. One time I was trying to cut a board in half with a knife. Louise told me that it was dangerous, but I insisted that it was safe, the tussle over the knife got intense but she was able to succeed. She was right I was in no condition to cut a board or anything with a knife. The steroid medicine was making me act "strangely." I was like this for about a month or so and I believe that I was driving my daughter and son in law to the crazy house. Finally, I was felt better and started crocheting again. Thank God!"

The Crochet Lady stands up slowly, reaches into her red crochet bag and put her work on display. She had afghans, doilies, baby booties, blankets and mittens just to name a few. The ladies at the center were in awe. Some of the ladies did crochet work but nothing in compared to the Crochet Lady. She was truly an artist.

CHAPTER EIGHT

The Demise

The Crochet Lady slowly walked into the kitchen. She always insisted on making her own breakfast. It usually consisted of a fried egg, a bagel with strawberry jam, strawberry cream cheese, coffee with ten teaspoons full of sugar and an assortment of pills for every ailment you could think of. Today was a good day for she had finished all her orders to be sold at the center. Her walker is loaded down like a mule taking goods down the trail at the Grand Canyon. Mike has this big smile on his face and just shakes his head. What a sight to see! When she arrives at the center the clients surround her and receive their pieces of art. They are all so pleased.

Time had passed and after lunch she continues her corrido... Then back to the orphanage...

Pantalion was now a year older at the orphanage. He is taller and stronger and becomes Mr. Villani's best wood shop student. When Mr. Villani isn't able to assist the other students Pantalion steps up and assist the students. Mr. Villani has total trust in Pantalion's skills. There were times when he would allow him to teach the class. Before students were allowed to use the power tools they had to pass a safety test. Pantalion would administer the tests to the new students.

On the contrary Billy was not a very good woodshop student. He would be yelled at by Mr. Villani on a consistent basis. His projects were never finished on time and were used as examples of, "what not to build." One day Billy refused to wear his safety goggles when working on the lathe. Pantalion, kindly reminded him to put on his goggles. When the teacher turned his back, Billy threw a block of wood at Pantalion and hit him on the back of the head. Blood gushed out from the wound. Pantalion grabbed the back of his head and looked at his hand that was stained with red. He quickly turned and saw that it was Billy. Pantalion reacted and looked over at Billy with the meanest look. He was ready to battle.

"What are you going to do about it, panty waist?" taunted, Billy.

Pantalion faced his adversary and rushed him as fast as he could. He lowered his head and tackled him like a linebacker. Billy's head snapped back and hit the brick wall behind the lathe machines, leaving a blood stain on the wall. Then with one arm and fist he hit Billy with a barrage of punches. He pounded Billy's stomach with such ferocity it made him keel over. Then Billy got up off the ground and faced off with his nemeses. Billy raised his right hand and threw a punch which was caught in midair by Mr. Villani.

"Now boys you'll have to go to Mr. McFarland's office for discipline. Don't you guys know how dangerous it is to fight, let alone in a workshop with power tools running all the time?"

"Yes Sir Mr. Villani, I'm sorry. I'm so sorry," pleaded Pantalion.

"I'll kick your ass, red headed Mexican panty waist, now you are in trouble with me you little wimp, no one gets away with this, no one," threatened Billy.

The class of students stood up and drew their attention to the exiting boys. With loud applause and cheers the students chanted, "Andy! Andy! Andy!" They were so delighted to see Goliath go down at the hands of 'Andy the Giant Killer.' They were so happy to see Pantalion standing up to this big bully.

The boys entered Mr. McFarland's office with fear because if you went to his office it usually meant that you'd be introduced to the "Board of Education." "The Board of Education" was not a committee of people, but a wooden paddle with numerous holes drilled into it. The holes would make it easier to move faster through the air. The hit on the buttocks would create a smack noise that could be heard throughout the school.

"Ok boys, now you know it's against the rules to fight here at the orphanage. Now, let's hear both sides of the stories. Let's start with Billy."

Billy gave his version of how the fight started and when he was hit from behind. Pantalion just thought that it was a bunch of lies and only hoped that he'd get only one swat.

"Okay Billy you've been in here too many times so let's go into the next room to get your swats."

Mr. Mc Farland grabbed the paddle and took Billy next door into the next room to administer the punishment. All you could hear was yelling, crying out loud and the smack of the paddle over and over against Billy's fat flabby butt. One smack, two smacks, three smacks and then four. Billy was screaming for mercy and the louder he screamed it seemed like the paddling got louder and louder. Pantalion sat terrified in the room next door to the point that he too was crying. He only wished he was home in El Guique in his mother's arms. The screaming and the crying ended and there must have been silence for ten minutes. In the background Pantalion could hear whimpering and then a door slam. Mr. McFarland and another adult were conversing and all Pantalion could hear was mumbling.

Seconds later, Mr. McFarland entered the room and said, "Andy, because this is your first offense you will go back to class and continue to be a well behaved young man. I know that you were only defending yourself and the bully had to go down."

"I will try my best sir, said a relieved Pantalion. Thank you so much sir." He couldn't believe what had just happened. For sure he assumed that he too would be screaming for mercy. As Pantalion walked back to the woodshop he looked out the window and could see the police hauling Billy away. It was what Jessie had said earlier, it would be Billy's demise. You can only get away with picking on the innocent for so long. Later, they found out that Billy was sent to a boy's prison in Las Cruces where he would have to spend extra time for being a habitual offender.

Pantalion finally made his way back to the classroom where he entered a cheering group of very overjoyed classmates. All of the kids surrounded him and praised him for his feat. They all loved Pantalion at the orphanage and wanted to be his friend. He felt good that he was able to defeat the giant but at the same time, felt bad for Billy.

"Damn Andy, you are one good fighter," complimented Jesus.

"Oh it was nothing; I just defended myself, "replied Pantalion.

"You are going to have to teach me how to do that tackle move with your head. It was a really great move and *muy chido*, (cool!)"

Back at the center...

The patients at the center just marveled at the Crochet Lady's *memory*. She remembered every detail from her past and the story continues...

Three years had passed and now Pantalion was 14 years old. He was a head taller than before. Strong as a mule, he was becoming a man. Still very humble, shy and how he loved his woodwork. It was three months until he would leave the orphanage when he decided to build one of those beautiful cedar chests for his mother. He had Mr. Villani order the best 1x6 cedar boards. He cut them to length dowelled and glued them together to make the sheets, cut them to the specified size, added little squirrel figures to enhance the beauty and made it open up with a door that could be sat on as a chair. This took about a month to build. Mr. Villani would help him with all the detail and sanding and after a month it was a finished product, a work of beauty.

The time had come. Three quick years had passed and it was time for Pantalion to go back to El Guique and reunite with his parents. On his last day Mr. McFarland and staff were very sad to see him leave, especially Jessie his best friend. Oh, of course, how could I forget his mentor and role model Mr. Villani his shop teacher. Also at the farewell was Mrs. Pendleton. She was holding a cloth like bag in her hands. Pantalion took the bag and asked what it was and she responded, "a nap sack for in case you have to go on a journey."

"Remember Andy, life is a journey. Someday you'll have to use this to carry life's valuables and dreams."

He had been waiting for a couple of hours when finally he could see his grandfather's 1925 red Ford pickup driving up over the hill to the complex. It was his grandparents the Garcias with no mom, dad, nor brother and sisters. He was excited to see them and asked, "where's my mom?"

There was a long pause of silence and his Grandfather said hesitantly, "Your mom stayed home and Enrique threatened to kill you if you were to come back."

"But why, why?" asked Pantalion.

"The entire community in El Guique was very upset with your father for what he did to you and didn't want to do any business with him anymore, which made him very poor for 3 years, said Grandpa Garcia. He blamed everything on you and said he would kill you if you would come back home."

Pantalion was very sad, embraced his grandmother tightly and asked, "Can I live with you and Grandpa?"

They paused for quite a long time and finally said, "He said he would kill us if you were to move in with us too." So its better if you move in with your Tio Juan who lives in Chimayo, you should be safe there.

Pantalion looked down at his cedar chest, sighed and said, "Grandma you take this and keep it for a while and then give it to my Mother for Christmas. Please tell her nothing until father has settled down. She doesn't have to know, but eventually let her know that it was a gift from me."

"Great idea," she said.

The Crochet Lady

It was another day of rejection and it seemed as if he was a total failure. He was only 14 years old and certainly had to grow up fast. Then the three were off to Chimayo to visit Tio Juan and Tia Ana. When he arrived in Chimayo he saw the beautiful Santuario. It was small Church which was known for its healing holy dirt.

"Let's stop Grandpa, so I can get some of this healing dirt to heal my broken heart? He asked. They say that the hole refills itself because so many come from all over for this holy cure. How can that be Grandpa?"

Then Grandpa Garcia shared the story he heard.

"Well mi hito, many say that the Penitentes, known as the Brotherhood of this area were praying out in the valley, on Good Friday Evening. Because Good Friday is when Jesus Cristo died, the men were in deep prayer. One of the Penitentes saw a light move in the distance. When the light was found, they discovered that it wasn't a light but a cross buried half way into the ground. The Brotherhood felt that it should be sent to another *Morada* (chapel) which was ten miles away. As the story goes, the cross made its way back mysteriously to the same place where they found it on three different occasions. The Penitentes then concluded that it was a message from God. So that's what we have here a small shrine with Holy dirt known as El Santuario."

"Grandpa you are so smart, how do you know so much?" asked Pantalion. "I too am a Penitente, a member of the Brotherhood of Jesus Christ. I believe!"

So at Pantalion's request his Grandfather pulled over and they went into the church. To his amazement he saw a huge dug out hole in the floor area. This is where people from all over the world would end their pilgrimages and dig out and take this holy dirt for miraculous medicine. On the walls were many crutches and wheel chairs where people with walking disorders would be healed instantaneously therefore they'd leave their crutches and wheelchairs behind. So Pantalion dipped his hands into the hole and gathered the dirt, unbuttoned his shirt and rubbed the dirt over his heart and asked God for forgiveness if he'd done anything wrong in his life. His grandparents did the same and then went upstairs and kneeled on a pew and prayed a Rosary to the Blessed Virgen de Guadalupe so their grandson would find peace in his heart.

Time had passed quickly and it was now sundown. They left and continued on the road to Tio Juan's. On the way there, Pantalion could see all the beautiful trees and acres and acres of planting fields. When they arrived at Tio Juan's his younger cousins, who were children came running out shouting with joy to see him. Everyone loved him. He was always so humble and respectable. After a time of conversation, Tio Juan said that he could live with them until harvest time is over. They needed Pantalion to help in the orchards to pick apples and do other small chores around the ranch. What was so nice was that his mother could also go to visit him in Chimayo. The best time to visit him would be when Enrique was off grazing the sheep in the mountains. She could be in Chimayo praying and visiting the Santuario and most importantly visiting with her son. This happened very seldom, but when they did meet they were so happy to see each other. Who can describe the love between a mother and son? The farewells were very hard for Pantalion and Roseana his mother, but they stayed hopeful until their next meeting.

Pantalion stayed at his Tio's ranch until the end of the harvest time. Then one day he was asked to leave because Enrique, his father found out that he was at his wife's brother's house which enraged him. So his journey continued. He was destined to move to his cousin's ranch in Chama, New Mexico. Chama was 120 miles from Chimayo. This was the beginning of his long, long journey to the north. The nap sack was a very vital gift from the orphanage. He was able to pack his clothes and stuff food enough for a week with apples, tortillas and other staples. Monetarily, he had a pocket full of change. It would have to be his will to get away from all the turmoil from his father.He calculated that the journey would take approximately a week to do. He prayed a rosary with his Tio and Tia and off he went, working his way to the Espanola to the west.

CHAPTER NINE

The Journey to Chama

Pantalion had heard many interesting stories about Chama, New Mexico. This is where his Primo Augustine was a sheepherder for the McClellan Sheepherding Company just north of Chama. Pantalion had some experience herding sheep with his grandfather in El Guique and thought that this was where he'd settle down, find a wife, and have a family.

Chama would be his sanctuary, his safe hiding place. But to get there he was on his own. Tio Juan pretty much washed his hands of the whole debacle and didn't want to have anything to do with the problem.

Now at age 15, he would be a man of his own caliber. There would be no more tears, no more fears, just a lot of heart to discover who he was as a man. It was his odyssey.

Back home...

After a long day at the center, the Crochet Lady arrived home and went directly to bed. She realized that her green doily was missing. She struggled to get herself out of bed and began to walk to the front room in search of the doily. She rummaged through boxes, cupboards and more boxes. She then remembered that it was under her pillow on the bed. So she'd wander off to the bedroom where she'd started her search. She then had a dizzy spell and fell to the floor. She got up quickly, thinking that no one would hear the fall. I ran to the bedroom and there she was sitting on the edge of the bed as if nothing had happened.

She looked up and said, "nothing happened, I'm ok and you need not worry."

"Maria you know how important it is to use your walker! God forbid one of these days you're liable to hit your head so hard that there will be all sorts of problems."

She tried to listen, but couldn't hear a word I said. Then she sits up, grabs her green doily and her walker and off she went to her favorite chair in the kitchen.

The kitchen table was her favorite place to crochet and work her arts and crafts. Her hearing was pretty much gone by this time we had to communicate with her writing a white board and dry erase marker. We would write her messages on the board and if the messages were written too long she'd only look at a couple of key words and misconstrue the whole message. I just wondered and couldn't imagine what she was going through. She had troubles with walking, seeing and hearing. But nonetheless, she always tried to fit in and be a part of any group. Most people were fooled and thought that she understood what they were saying, but most of the time she was pretending.

When she fell, we were instructed by her Doctor to let him know. Falls were usually indicators that she was in her final years.

Back at the center the next day…

When the Crochet Lady arrived at the center, there was no waiting. All the clients surrounded her and of course the story continues…

To get to Chama, Pantalion first would have to get to La Puebla, to Espanola, to Mendanales, to Abiquiqu, through Carson National Forest and finally to the McClellan sheepherding ranch in Chama.

There was no correspondence with Augustine, who knows, he might not be living and working there, but nonetheless he took a chance. I just couldn't imagine a young 16 year old taking a journey with no idea what might happen. So much at risk he could die and no one would ever know where he was. But, anything was better than the threat of death at the hands of his father. So that morning he ventured off and took the road to Espanola. Espanola where many knew his family and friends lived. El Guique was a very short distance away. What might he encounter?

Everyone knew the Sisneros family in Espanola. Pantalion would have to tread lightly in Espanola just in case someone would recognize him and spread the news to Enrique. The eight mile walk to Espanola was a true test to what was to come. Many of the neighboring ranches had dogs that were loose. Thus, Pantalion would have to carry a big stick just in case one would attack. Just before reaching Espanola there was the small village of La Puebla. When he approached the village he was greeted by four very large mongrels, probably a Rottweiler mix. Pantalion knew that he was in for a fight.

"Okay this looks like first obstacle on the way to Chama," he murmured to himself. Back off, back off you stupid dogs."

But the dogs drew closer and closer showing their big yellow teeth with dribble and drool spitting out. The biggest of the four came forward when Pantalion cocked back the stick and swung as hard as he could. On the first swing he missed. On the backswing he caught the mongrel on its jaw full force. This only got the wild beast more upset and was ready for a fight. The force of the swing put Pantalion on the ground. He scraped his knee and then he felt a pinch on the back of his right heal. The smallest of the four canines had Pantalion's foot in its mouth. He shook the dog off and stood up quickly. He continued to swing his stick until the big Rottweiler mix leader was knocked out. Then from a distance at one of the houses in the village came out a man who shot his rifle into the air.

"Get out of here, *perros pendejos*! (stupid dogs!)" Shouted out a villager from a distance

"Gracias senior, you saved my life, I thought for sure I would be seeing St. Peter at the Pearly Gates," said an appreciative Pantalion.

"My name is Juan, Juan Quintana; we've been trying to kill those four wild dogs for the longest time. They've been killing some livestock in the area and it looks like you've killed their leader. Congratulations, you sure know how to handle a stick."

"Thank you sir, my name is Pantalion Sisneros. I'm on my way to Chama to work with my cousin at the McClellan sheep ranch," explained Pantalion.

"Chimaco, kid it looks like you've got a long way to go. We better go inside and tend to those scrapes. Tienes hambre chico, (are you hungry?)"

Only four miles from his Uncle Juan's ranch Pantalion found a very empathetic man who heard his story and genuinely cared to his wounds. Juan fed Pantalion because during the fight with the dogs Pantalion had lost all of his staples for his journey to Chama.

"Gee, Pantalion it's a miracle that you're alive." You're on your own, and I can only allow you to stay overnight and Ill drive you to Española in the morning."

"Thanks Juan. I owe my life to you."

"Tomorrow I will take you to the interstate, remember to take the road to Chama which is northwest, if you take the road to the right you will be heading to Taos and will have to go through Pueblo Indian territory," instructed Quintana.

Pantalion insisted that Juan drop him off at the edge of town. When Pantalion arrived in Espanola he knew that he had to be careful. His biggest fear would be running into Enrique.

As he walked north on Main Street from a distance, he could see the red pickup truck that his father parked in front of the local cantina. Suddenly, he broke out into a sweat and his body shook uncontrollably. He altered the course and went back and around the cantina. To his amazement he saw two men conversing. One happened to be Enrique his father and the other was Mr. Gutierrez, the Dry Good store owner. They were taking down shots of whisky. Pantalion so discreetly made sure that he stayed low so he wouldn't be seen. He rolled under an old rusted pick up and listened to what they were saying.

"Senor Gutierrez, my life has been a mess since Pantalion has left our family, I truly believe that he is not my son, explained Enrique. He better not show his face around here or I'll be sent to prison for murder. He has really ruined my life. My wife doesn't speak to me, we have no affectionate relations and all of the relatives hate me."

"*Andale* Enrique, you are a *Penitente* with the Brotherhood, a Roman Catholic and you believe in Jesus and La Virgen de Guadalupe. Now is the time to ask for forgiveness and that will set you free. There is nothing that you can do about the past, but I do believe that he surely looks like you," assured Senor Gutierrez.

At that very moment Pantalion felt a tear roll down his face. Then something slithered over his legs and he knew that it was not a mouse but a rattlesnake. Immediately, he rolled from under the truck and stood up. From that point Pantalion looked in the direction of the two men and for a split second locked eyes with his father. He then ran as fast as he could down Main Street to find a hiding place to escape the pursuit of Enrique. He passed building after building and found his way to an arroyo (creek,) aligned with rows of sagebrush. This is where he spent the night, sleeping amidst the bushes and smelling like sagebrush.

The next morning he crawled from under the brushes and went north where he went into a small grocery store where he bought a bottle of pop and a bag of jerky. This was the last of his money. How would he survive this journey?

For the next two years Pantalion lived like a vagrant, traveling from village to village asking and begging for food. At times he worked doing small jobs. There were even times where he was picking out food in garbage cans. It was a sorry sight to see.

The 120 mile trip became a disaster for he found himself on to the road to Taos not Chama. As he experienced being down trodden, his will to live

The Crochet Lady

was at an end. There were thoughts of suicide. Ending his life would solve the problem. But somewhere deep inside he had the will to live.

"If I could just make it to Taos, someone might have the heart to open their home and give me a hot meal to eat and a place to sleep. I could find my way to Chama. In the distance he could see a herd of cattle gathering at the water hole to relieve their thirsts. He said to himself, "I am so thirsty, if I could just drink the water from that puddle I can make it to Taos."

Then from the middle of the herd he heard someone shout, "*Orale*, (hey) Chico what are you doing? Are you out of your mind?"

"I'm sorry sir, I'm suffering from thirst and I need to make it to Chama," explained Pantalion.

"Chama, sorry boy, but you are on the wrong road. You need to take the next road west as you get into Taos, instructed the cowboy. I'll take you to the priest in Taos. He'll know what to do."

Hence, on the ride to Taos was surely a relief to Pantalion's much worn feet, shoes and starving body.

Hereafter, the story continued as Pantalion shared his story with the Cowboy.

"Damn, you are quite the survivor and damn!"

"You are one stinky Mexican who needs a bath like a year ago. Whew!" emphasized the cowboy.

Pantalion laughed. The cowboy could see sadness in Pantalion's eyes. The cowboy had a very kind heart and figured to get help for the Lion's Gate he'd take him to Padre Joaquin at the church called Our Lady of Guadalupe in Taos.

"Padre Joaquin will know exactly to do with you. He is a very wise old *Padrecito* (priest) who knows a lot of people who can help you.

Therefore, Pantalion also known as the Lion's Gate, stayed at the Church doing odd jobs around the church's compound. He was fed and clothed. He was Padre Joaquin's assistant at all masses, funerals and weddings. Some people believed that one day he too would become a priest. But there was always that calling for him to go to Chama and work with Augustine at the McClellan sheep ranch.

Pantalion was 20 years old now and felt that it was time to leave this great experience in Taos. Through correspondence, he had Padre Joaquin write a letter to the McClellan Sheep company to see if Augustine lived there in Chama. Subsequently, Pantalion was not well versed when it came to letter writing. The Priest was highly educated; it would be the smart thing to ask Padre to write the letter. Before he would go to Chama, he had to be assured that Augustine was there in Chama. Two weeks letter he received his answer.

Dear Pantalion,

It is great news that you are alive. There have been numerous stories of you dying by some kind of tragedy. Yes, I'm here at the ranch and Mr. McClellan wants to hire you to herd sheep. I told him that you had experience guiding the flocks of sheep en La Sierra. He is excited to meet you. He's giving you a week to get here. If you're on foot make sure that you jump the freight train outside of Taos. It will take you across the Rio Grande Bridge Royal Gorge and then continue until you get to Chama. When you arrive at the depot at Chama, just ask for Augustine Muniz. See you then Primo.

Sincerely,
Augustine Muniz Sheepherder

Some various craft art made by The Crochet Lady when she no longer had the ability to crochet

I guess he could have bought fare to get on the train to Chama but did not have the money to pay for it…

Padre Joaquin and Pantalion walked to the street outside of the Church's compound and gave their farewells and were certain that one day they would meet each other again. Tearfully, Padre Joaquin had Pantalion get on his knees and gave him a blessing and asked him to go daily under a tree to pray a rosary if he couldn't find a church to pray in.

So the saga continued…

Pantalion stood bright eyed and ready to take on the world. His thoughts were of the past, the dog fight in Puebla, the snake slithering over his legs in Espanola and the terrible hunger and thirst he had on the wrong road to Taos. Of course there was the court decision, the fight against Billy and the great food at the orphanage. It was four years since Chimayo and finally he would find his way to his destination, Chama, New Mexico.

In order to get to Chama, he had to get to the train. He had to walk west outside of the city, which was a desert filled with cacti and sage brushes. It was mid-morning and the sun rays were increasing in temperature. In a short distance he could see the railroad tracks. The railroad track spread out east and west. He waited several minutes when another commuter came out from the bushes. He gave a quick introduction of himself and gave Pantalion instructions on how to jump onto the moving train.

"I'm not sure on how to do this, you see, I only have one working arm" replied Pantalion.

"It's easy my friend, just run alongside the train and then reach up and grab the handle and pick yourself up into the open box car," said the man.

Then and there was the train, right in front of them. There were many open box cars that could be jumped into. If he missed his chance with one there was always another box car. Then the man instructed, "Now watch, and do what I do."

This would be a scary thing to do with one arm let alone two. He watched the man carefully and memorized every step. He murmured to himself, run fast, grab the handle and pull yourself in, jump into the open boxcar.

Now it was his turn. He ran as fast as he could, reached up and went to grab the handle with his good hand. Suddenly, he stumbled and could only get the one hand on the handle. It was a struggle. His feet dragged on the ground. He could feel his feet hit every rock on the track. Then with full force with that one hand he pulled himself up into the boxcar and rolled head first into the unit. To his astonishment he accomplished the task. In the background he could hear laughter.

The men inside the car were entertained by the look on Pantalion's face.

He was covered in sweat and a loss of breath which made him look like a boxer after round 12. He was perplexed and embarrassed from the men laughing from within the boxcar. Inside the boxcar were approximately 10 other men sitting up against the walls of the compartment. Most of the men were vagrants too. They were traveling from city to city making a living here and there by scavenging. They too were travelers on the train from Taos to Antonito and eventually to Chama.

"Chico, you did a great job by using only one hand to pull yourself up, praised one of the men in the compartment. Where are you headed and what happened to your arm?"

Pantalion looked up in embarrassment and said, "I don't like to talk about it because it's a wound that I'll have to carry on for the rest of my life.

"No disrespect Chico, but everyone on this train has some sort of physical disability or as you put it, a wound. Thomas over there has only one eye. Jack was stricken with polio and has one leg smaller than the other and Juan was born with both ears missing. So Chico we all can relate," added the vagrant.

This made Pantalion feel better. He shared with the others his story and all the trials and tribulations he'd gone through. It was his dream to work in Chama and shepherd the flock.

"Well Chico you know that we have to travel across the gorge at the Rio Grande River. Some people say that the gorge is haunted. There is a Puebloan legend that says that the gorge has evil spirits within it. Many people have been driven to its rim and have thrown themselves to its depth into death, suicide. So, when we start to cross the gorge it's a good idea to go to the center of the boxcar and embrace each other so that everyone will survive the crossing without being tempted to jump to its depth," said the man with one eye.

Then the man with no ears said, "You will know when the gorge is asking for a body. It's when the gorge sings out a song. That is sung in cadence with the rolling over the tracks, bumpity bump, bumpity bump, bumpity bump. The noise will be so loud that you'll have to scream at the top of your lungs to fight off the evil spirits."

Pantalion was shaking uncontrollably and started to pray the rosary. He mouthed out the words *Santa Maria Madre Dios*, (Hail Mary Full of Grace.)

"We are getting close so let's all gather in the middle of this boxcar and hold hands," said the man with one eye.

Hence, they all gathered in the center of the boxcar, held hands and began screaming as loudly as they could. It was intense. In a matter of five minutes they had crossed the gorge safely and were all accounted for. There was complete silence for a moment and then the group busted out with a roar of laughter. However, Pantalion never told them about the figure of the bearded man that he had seen at the opening of the boxcar calling him to the opened door. For the rest of the night he slept with one eye open and fingered the beads of rosary that Padre Joaquin had given him.

The next stop would be Antonito, Colorado and then on to our final destination, Chama, New Mexico. In order to get to Chama he would have to go to Colorado first and then back across the border into New Mexico. The stop at Antonito was a half an hour layover and the men found time to go to the stop and beg for food from the tourists who were vacationing in the San Luis Valley.

CHAPTER TEN

A Proposal

Back home...

Louise tried to make it easier for me to communicate using the white board and dry erase markers. She also set the television to close caption to watch my favorite TV program, "Wheel of Fortune." She purchased an emergency system through the telephone company which was also set up for closed caption reading and communication. I would wait by the telephone patiently to wait for a call, but not too many people would ever call me. When I lived with Louise and Albert for 4 years, I probably received 4 calls. It was so sad for me, because I missed all of my friends and relatives greatly. I just wanted to hear, "hello, how are you," or see it on the screen once in a while. Sometimes my days were so sad and I felt very, very lonely.

The Crochet Lady would sit by the phone for hours and yes, to no avail no calls. Not too many came to see her in her final years and when they did it was always a treat for her. In all of this she pondered and some days her depression was spread to us all. It hurt to see her so sad and melancholy. But one day we had a big surprise from the family, a surprise party when she turned 90. It was so pleasant and nice to see everyone there. Connie came from California who was named after her mom Constancia. Alonzo's children were there, Danny, Diego, and Damian, Andy her eldest son with his two daughters Jody and Lorie and their spouses and children. Her sister Catherine and her children just to name a few... The party was so nice that it came at a time when she needed some hugs and kisses.

Back at the Center...

"Maria, was there ever a proposal of marriage?" Asked Alvie

That's funny because when I was 17 years old and bored to death with my boring life, we had been working, picking crops in Monte Vista. An older man, around 35 years old and a good friend of Tio's asked for my hand in marriage. He had 3 children and was like us, dirt poor. My step dad was willing to marry me off to what I called a *viejito* (old man). He was very handsome but I wanted to share my life with someone more my age. He had written a letter to my parents and we used the old tradition of the "pumpkin of rejection," a big orange pumpkin. The tradition goes as follows, a letter is written to the parents of the daughter. The parents would review the letter. If the proposal was accepted the girl's parents would write back saying "yes" and if the answer was "no," a big orange pumpkin was delivered.

There were many good looking guys in Capulin and the San Luis Valley but I was still waiting for my White Mountain. I prayed every day that one day he would walk over that mountain and I'd know immediately that it was he. In some people's eyes I was at the age to get married. Age 18 was on the edge of being an old maid, but I didn't care I knew that one day my Mount Blanca was coming over that mountain to rescue me from this monotony. So my prayers were many and it seemed that the Lord had different plans for me.

CHAPTER ELEVEN

King Henry the Eighth

King Henry the VIII was the King of England in the 1500's. He always wanted to have a son carry on his name and be the next King of England. There was only one problem, all six wives would not conceive a son and gave birth only to girls. King Henry the VIII was so upset that he couldn't have a son and blamed his female counterparts. He asked for permission from the Pope to divorce his first wife Catherine of Aragon and this was denied in the Holy Roman Catholic Church that he separated from the Church and started the Church of England where he made the decisions on the new religion which allowed annulments. So every woman after that was blamed for not bearing him a male child.

I finished my schooling at the Gomez school k-8. If we wanted to continue our schooling in senior high I'd have to go to Monte Vista. Now that wasn't going to happen. Mom and Tio wouldn't spend the money for me to go there. They thought that it was a waste of money. To tell you the truth I don't even think they thought about it. Everyone in those days believed that an eighth grade education was plenty. It was just one of those expected things. You'd go to school until you were 14. Possibly several years later a young man would ask for your hand in marriage, you'd have children and live happily ever after. But that didn't happen for me, for I stayed at L'Ojito and worked and worked and worked.

Tio had great dreams for all of us. Yes, it was his dream for me and my three half-sisters to work, work and work. My sisters were Catherine, Orelia who born in 1930 and the baby of the family Emily who was born in 1934. But ironically they never got their son. There would be no son to carry on the Lucero name. It reminded me of the story of King Henry the XIII.

I must have been 17 years old when Tio came to the family with a new idea on how to make money for the family. At an auction, he outbid the others by purchasing an airplane engine and a huge saw for cutting big timber.

"Guess what we are going to do?" Tio asked

"Fly airplanes?" Catherine asked.

"No, don't be so silly *hita*, (daughter) we are going to start our own sawmill business here in L'Ojito, right in our own front yard. I've gotten permission to saw the big timbers just above us in the high country. I can sell to the local people fire wood. The big planks we'll sell to the lumber store in Monte Vista. I've also negotiated a contract with the lumber company in Monte Vista. This definitely is a great opportunity! Exclaimed Petro

Constancia was very concerned about the new venture. It was hard to establish such a business because it would require much revenue and man power. There were other men who could help with the business. In L'Ojito there were many men without work. In Capulin, just to the east, there were many veterans who needed jobs after WWII who could also help.

"What can I do? I want to help out." Asked Emily

Petro slowly put it all together. He worked many hours a week and established his new business. There was no stopping him to achieve his goal.

"Well first of all we need to hire a mechanic who could put the saw and attach it to the airplane engine. Your primo cousin Ronald from Capulin who just got out of the service has lots of experience working with engines. He should be cheap and give us a good deal. He said he'd do it for a case of beer, but I'll buy him the case plus the money." informed Tio.

"Daddy do you think we will be able to go on trips with the money?"

asked Emily.

"Maybe," said Petro

"This will be great for our income. We will be able to buy some more livestock and some needed things for the house, reminded Constancia.

"We have much to do, I've written down a list of chores in order to make this business work," instructed my stepfather.

So for the next seven years I became a lumberjack for the Lucero Logging Company, proprietor Petro Lucero. At first, many from Capulin wanted to work at the mill. Tio didn't pay his workers much and worked us like slaves. Little by little, men would quit or be fired. After a year or so, the labor force was just us, mom, Catherine, Orelia, Emily and I.

I became as strong as a man and in some cases stronger. I could lift big tree trunks and load them into the trailer or what we called a sled. The road to the trees was very treacherous with ruts and in some cases washed out by the rain. Many times it was just Tio and I cutting and lifting and then down to the sawmill where we would unload the trunks. It was a very monotonous type of work and in time I was developing big muscles. My sisters too were very strong and in good shape.

Men and women working at the Lucero Logging Company in L'Ojito Saw Mill

When the trucks were out of commission the horses would sled timbers down to the saw mill

CHAPTER TWELVE

The Rifleman

If times got bad at L'Ojito in terms of shortage of food, there was always the forest to the west of us. The forest had herds of deer and elk. We never ran out of fresh venison to eat, especially when a visitor with all the skills of a sharp shooter arrived to feed our thriving community. He went by the name Johnny, yes Johnny Garcia.

There he was on his mighty steed with is 306 over his right shoulder, a black Cowboy hat, and a lariat tied to the horn of his saddle. He was a marksman at the young age of 15. Handsome to say the least as all the ladies old and young would turn their heads to take several looks at such an Adonis.

"Who is this young man?" asked Catherine, "He's your distant cousin and you better stay away from him," declared Constancia.

"But mother, he is so interesting looking, I can't explain it. He is so handsome," said Catherine as she grabbed the comb to take out her tangled hair. "Mom, how do I look?"

"What do you mean, how do I look?" responded Constancia.

Actually, I thought that this Johnny was very good looking too. However, he was too young for me. I still prayed that my Mount Blanca was on his way to rescue me from this nightmare of a life.

Johnny was put to work immediately at the saw mill. The job was very grueling. To make it harder Johnny couldn't keep his eyes off Catherine. Petro would get so upset on two accounts; one because production was slowing down, and the other, his daughter was falling in love with this vaquero with a 306 rifle. The two would run off and they couldn't be found for hours. In desperation, Tio got smart and assigned Johnny to all hunting chores.

Johnny became quite the huntsman. He was a marksman with his thirty odd six caliber rifle. He was so good that he usually would bring down 2-3 deer and occasionally an elk. With the power of his rifle, Johnny became such a sharp shooter that people would hire him to bag them a deer or two.

The deer meat was so delicious. We'd hang it upon a stringer like (beef jerky) *carne seca*, along with dried zucchini and *chicos* (dried corn.) On the day of the kill my mother would make a nice venison roast. Wow! What a feast and when the potato harvest came around the feast was enhanced with fresh mash potatoes, *chile colorado* (red chili) and *frijoles* (beans) with *chicos* and of course flour *tortillas*. My mother was such a great cook. For Christmas she'd make the best empanadas made with deer meat. Deer meat was a staple in L'Ojito, shot illegally but the authorities turned their heads because most of them were poaching themselves.

Oh, what a crazy time! Just to think a year earlier Mrs. Garcia from the Gomez school wanted Catherine to move in with her in Monte Vista to attend High School. Mrs. Garcia was so impressed with Catherine's aptitude that she felt that she could finish her schooling at Monte Vista and then be her replacement at the elementary school when she retired. Of course Tio Petro said no, because he needed her to work at the saw mill.

Mother would get very upset with my sister because she was constantly talking and flirting with this rifle man from Pagosa Springs. Things worsen but there was not too much Tio and Mom could do. In a matter of time the two wanted to get married at a very young age of 16.

About two years after the wedding, Johnny and Catherine built a two room cabin behind the main house where I lived. As expected they started their family immediately one year after another with two baby girls Paula and then Margaret. It was a trying time for my sister and brother in law. Mother would cry and only hoped that something good would eventually happen to the help the young couple.

Johnny felt that he couldn't support Catherine solely working at the sawmill. He needed more. Like Regino he couldn't get along with his suegro (father in law.) Therefore he ventured off to the north to find a more lucrative way of surviving. Many also believed that Johnny would not come back to Catherine.

Oh how I felt for my sister. Johnny was gone for a while to the north and promised to come back to L'Ojito to get Catherine and the girls. She was pregnant with her third child and only could pray for her man to return to L'Ojito.

"Oh, now that husband of yours is never coming back, the only thing that he's good for is hunting and getting out of work here at the saw mill," declared Petro.

Catherine was 19 years old, two daughters and one on the way. We all were concerned that we would never see him again. I prayed that Johnny would return to L'Ojito. Then my prayers were answered.

Later through correspondence Johnny wrote Catherine a letter saying that he was lucky to get a job in the winter working at the stockyards in Northern Colorado for a big meat packing company. Time had lapsed; Johnny had worked in Northern Colorado for about a year. We still didn't hear word when he was coming back. Catherine had an undying faith that he'd return.

CHAPTER THIRTEEN

Search for Gold

About this time in history, Pantalion was an experienced sheep herder for the McClellan Sheep Herding Company in Chama, New Mexico. He had never been more proud of what he had accomplished. In charge of a flock of 200 sheep, a beautiful quarter horse named Silver, a 30/30 Winchester to run off bears and coyotes and his trusty Australian Collie mix named Pokey. Some people considered him to be a self-made man. The McClellan's fell in love with him and allowed him to live in the bunk house next to a home where Augustine and his family lived.

It had been several years since Pantalion would take the sheep to the sierra over the border from New Mexico into Colorado. The best grazing land was that stretch between Chama and Capulin. But the journey could be very dangerous because of the wild predators. Pantalion always had to be at his best, managing the flock like the Good Shepherd. If one sheep would run off, he'd have to keep the flock together with the skillful Pokey and go off to find the stray.

His trusty dog Pokey was well trained and could guide and separate the sheep at Pantalion's commands. His commands were, andale Pokey, which was the command to take the sheep (forward at a fast pace ahead.) Arriba was his command to take the sheep (up and over) the hill. A la derecha and izquierda were the commands to move (left and right.) His favorite command was a sharp quick whistle which meant it was chow time.

After about the third year of sheep herding, Pantalion got word from the town folk that there was lost French Gold in the area where he had taken the sheep to graze yearly. Augustine was also interested in finding this lost French Gold that everyone was searching for. So, they planned to work as a team to find this lost treasure.

"You know the story we heard was that these French Canadians, buried the gold just west of Saguache in and around "Treasure Mountain," said Augustine with a smile on his face. "There are many stories about this lost treasure but this one is the most believable. This is where we'll graze the sheep."

"I surely hope that that you are right primo," replied Pantalion. "I suppose that you also need to know that the gold is haunted. "Haunted? Why do you say that?'

"Well, when the French went back to retrieve the gold from the buried spots, they couldn't find the gold. Then the Utes ran them out of that area. So the mystery is, did the French Canadians retrieve any of the gold? The French Canadians went further south to where we are going to search for this mysterious gold, "explained Augustine.

"It's probably a bunch of *mierda*, (bull shit) told to you by someone who already has the gold hidden for them," teased Pantalion.

"No Pantalion, you are the one who is full of shit," retorted Augustine.

"Let me finish the story primo."

The Canadian Frenchmen buried the gold in the area where we will take the sheep for grazing and when the Indians drove them out they hid the gold in a bear cave so no one would find it. Years later, some of the surviving Canadian Frenchmen went back to the cave and couldn't find it. After a week of frustration the French ran into another attack from the Utes and were all killed.

"I guess that we are very lucky that the Indians are not attacking anymore," said Pantalion with a rolled cigarette in his mouth.

The responses from Pantalion were mumbled because he had a rolled cigarette in his mouth. Augustine still understood the garbled responses from Pantalion's mouth but just laughed to himself as he turned the reins of Cobra his horse to the left to avoid the cacti that was in his way.

"But you know the story doesn't end there cousin. This is what makes the gold haunted. Several years back a family was camping right above a little hamlet called L'Ojito. The area was flushed with vegetation. There were big pines and aspens. It was the perfect place to go camping and to the west was a lake where people could fish," explained Augustine.

"That sounds like a place where I could build a cabin and raise sheep and who knows maybe find me a wife and raise some little Pantalions." responded Pantalion.

"Ok Pantalion, now you are getting a little *poquito loco* (crazy)," said Augustine in jest.

"Will you please let me finish the story? Please, begged Augustine with frustration.

"Where was I before you interrupted me?"

"The family, camping! *Pendejo!*" (stupid,) reminded Pantalion.

"Okay, one day one of the girls was playing fetch with her dog. Upon waiting for her dog to return the ball, the dog just continued running into the forest. After several hours the dog didn't return. In desperate search for her dog she went out to find it. She finally heard some barking. The barking grew louder and louder as she got closer to a barking sound. She then looked down and saw what looked like a hole in the ground. Hesitant at first she finally cleared out the hole with her hands and looked inside. There he was. She knew that she needed a flashlight or candle to see what was inside that was blocking the dog from not coming out. So, she went back to camp to find her father who would know what to do. She led him back to the hole in the ground. The dog was eventually able to free itself from the hole. The dog was so happy to be out of the hole and the little girl cried out with joy. The father brought a flashlight and a candle and wanted to know what was inside the hole. When he flashed the light through the hole he noticed the light was reflecting off a rock on the other side of the hole. Then the father got a candle and placed it where the light came through on the other side. After he did this, he peered through the hole and saw three large bouillons of gold shining like the sun. "What could this be?" He thought. He was so bewildered but knew that he'd have to get the right tools to get the gold out. It was night and he knew that to get the work accomplished he'd have to come back in the light of day. He set his bearings in his head and marked off a couple of trees with yellow ribbons to make sure he would find his way back."

"Well, what happened, I sure hope the dog was found," said Pantalion with a worried look on his face.

Now you sound like the pendejo, the dog probably was okay, but it's the gold that needed to be found," responded Augustine. "Didn't you hear what I said? The dog was able to free himself, of course!"

"Here we go again, will you please let me finish the story," pleaded the elder cousin.

"The next day the family went back to find the hole in the ground. They searched and searched and to no avail ever found the hole or the marks left on the trees to find their way back to the gold," concluded Augustine. "To add, when the family got into town, the father spread the news about the gold he saw near L'Ojito. Then the mad rush was on to find the gold."

"What do you mean, "mad rush?" asked Pantalion.

For many years after, people from all over the world came in search of this "haunted" gold," exclaimed Augustine.

Pantalion looked amazed and asked, "Are you sure you want to search for this "haunted" gold?"

"Well Pantalion, you only live once and this might be our only chance to make it rich and maybe have a cabin or two before we meet our maker," said Augustine as he blessed himself with the sign of the cross.

"Here's a map of the general area where the gold was supposedly hidden.

The area that we take the sheep goes over where the gold was told to be," illustrated Augustine with a map in his hand.

The rest of that day was spent for gathering all the supplies. The two knew the area pretty well and just had to make sure that they would look in the three general areas where the gold was told to be.

The two "miners/sheepherders" were in search of gold and were determined to find it. Before the two took to their quest Mr. McClellan met with them. He was somewhat concerned about how many sheep would be taken to graze.

"Why are you two taking mining tools?" asked the boss.

"Don't worry boss McClellan your sheep will be safe with us and when it comes time to take them to the market, they will nice and fat," assured Augustine.

The two went off and traversed directly northeast approximately until they crossed paths with Dominique LeBlanc. Dominic was a local sheepherder who too had grazed his sheep in the same area Augustine and Pantalion were headed.

"*Que pasa vecinos,* (what's happening neighbors?)" Dominique greeted.

"*Hola mi amigo,* (hello my friend)" the men said in unison.

"I heard that you are going in search of the lost French gold. Is this true?" asked Dominique with a concerned look on his face.

"We sure are neighbor. We figure if we are focused and persistent we can find this gold and most importantly try to keep ourselves out of dangers way." said Augustine.

"I just have one word of advice, my friends, "Bigfoot," warned Dominique.

"Bigfoot, what in the hell is a Bigfoot?"

A Bigfoot is a huge looking half bear half man super monster that can tear your head off with no sweat involved. He goes by many names for example; 'Sasquatch,' 'Skunky Bill,' 'Booger Man' and of course 'Bigfoot.' It is said that he stands upright about 8-10 feet tall and can weigh up to 500 ponds. You never see it in the day because it has great camouflage capabilities. It hunts by night and loves to feast on stray sheep."

"Come on Dom, what are camouflage capabilities?" asked Pantalion.

The monster could be 20 feet from you and you wouldn't know he's there. He blends with the environment. If there are trees nearby, it looks like a tree. If there's tall grass, it blends with the grass," explained Dom.

"Oh come on Dom, you are just trying to scare the heck out of us," responded Pantalion as he grabbed his 3030 Winchester.

"Yea Pantalion, you better have that gun loaded at all times because they love to feast on wildlife and of course their delicacy, sheep," pointed out Dominic.

Therefore, the two *primos* (cousins) rode off and pushed the flock northward into Colorado. The trip was very treacherous, going up and over hills and valleys and through streams. The terrain didn't matter for the two shepherds were skilled at the task at hand.

"Primo Pantalion, do think he's just trying to scare the hell out of us?"

"Well Augustine it's getting close to the time where we have to round up the sheep and grub on some of those good tortillas and frijoles con chile verde your wife made early this morning."

Pantalion never answered Augustine's question and just continued doing his job; rounding up the flock, setting up the tent and getting the fire going. Augustine set up the chow and made up the interior of the tent with sleeping bags and cleaning materials. It was like a home away from home.

That morning the two men arose at 5am and ate a quick breakfast. Immediately, Pokey knew that it was time to push the flock. There were some stubborn sheep who didn't want to move to the next piece of pasture land so Pokey began biting at the heels of the stubborn sheep. Finally, the sheep started moving over the hill to the destination. It was quite a sight to see, all of the sheep moving in unison. Imagine a cotton field moving forward up and down in rhythm with its billowy cotton balls all in a row. If you put music to the movement you could have great big baile (dance.) Pokey just loved to walk on the backs of the sheep. Then with a quick short whistle Pokey would jump off their backs and circle the sheep so that not one sheep was out of line.

It was about three miles to the next piece of pasture where the sheep could settle down. This was the ideal place to bed down for the night. Sitting and relaxing before the campfire the two men rolled themselves a cigarette and reflected on the past.

"Do you think you'll ever forgive him?" asked Augustine.

"Who are you talking about," answered Pantalion.

"Well that's a no brainer. Enrique! Your father," yelled Augustine.

"Quiet, *pendejo*, you are going to spook the flock, snapped Pantalion. That's between Enrique and me. Oh, I don't know, possibly and now that I'm a man I don't think he'll even think about killing me."

"I like that forgiving attitude," said Augustine.

For the remainder of the night they shared stories about the past. "Augustine you never talk much about your past,"

"Cousin Pantalion, there's really not too much to mention about my life."

My life was similar to yours. I grew up in El Guique, just about a mile from where you lived off the banks of the Rio Grande River. We had several horses, cattle and chickens. My father had an orchard of wine sap apples. My mother died from pneumonia when I was 15. My father, your Uncle, couldn't support us all. So when I was 17 I enlisted in the Army. I would send money back home to my Papa to help make ends meet. I left to Germany and actually fought in the war. Here at home I left my girlfriend in hope that when I returned we'd marry and have a family. Her name was Wilminita. They called her Billie. She was so beautiful. You see Pantalion,

Angela was not my first choice. Billie was my first choice. When I was overseas all I could do was think about holding her in my arms. Her body was so soft and how I missed kissing her soft lips. Well when I came home I got the bad news.

"What was the bad news cousin?" asked Pantalion.

She was married. She was expecting a child and no one had the decency to tell me. I was so angry. I wanted to kill someone. I cried and cried. My life was over. I had to get out of El Guique or there might be a homicide.

"Cousin, who did she marry?" questioned Pantalion.

I don't think it really matters cousin. I think this all happened when you were in the orphanage in Santa Fe.

"Who was he?"

He was some "asshole." He went by the name of Garcia, a big time rancher.

"Hell with her!" "Angela has to be better!"

"Not then Pantalion, Billie was my heart and soul."

It took me quite a while to heal my broken heart. But God always has a different plan. It was about two weeks later when Mr. McClellan came to El Guique looking for sheepherders. Without hesitation I applied and got the job. Mr. McClellan has been so nice and good to me. He gave me room and board with a lot of opportunity.

"Pantalion, I always had you in my prayers when you were lost and searching for Chama," interjected Augustine.

"Thank you cousin, I owe it all to you."

"When I got to Chama I was so lonely and couldn't keep my mind off Billie. I was dying from heart break. Like I mentioned earlier, God had a plan. It was about a week later when I met Angela at a wedding dance in Chama. Angela is a distant cousin of the McClellan's. She was all alone in her life too. Her first husband was killed in WWI. I have to say that she caught my eye. As you know, she's a beautiful lady. Well that night we talked and talked about our pasts into the early morning. After a while we couldn't keep away from each other. She was the one who was meant to be Mrs. Muniz. God knows what he's doing. That's why you are here Pantalion. My prayers were deep with the Lord."

"*Gracias Adios* (thank God) and thank you Augustine," said Pantalion as he grabbed his rosary.

Then the two men went under a big tree and recited the rosary.

Later in the middle of the night, from a distance, they heard a howl'. It was not the sound of a wolf or coyote, but a loud deep sound similar to a bear that was wounded. Pantalion jumped out of bed and grabbed his trusty 30/30 Winchester rifle and his lantern to see into the dark night. Pokey started to bark unceasingly. Augustine too was out of the tent and had his gun cocked and ready for battle.

"This might be Booger, Augustine. The Bigfoot that Dominic was talking about. We better not shoot because it will cause the flock to stampede," instructed Pantalion.

Both of the men were afraid to go any further than 10 feet from the front of the tent. They decided to rebuild the fire and make the flames big to keep Big Foot at bay.

It seemed like the fire kept the monster away. After several hours the men set their sleeping bags in front of the tent next to the fire. Pokey settled down, which was a clue that the beast had fled the scene.

CHAPTER FOURTEEN

My Best Friend Alvie

"Okay everyone, it's time to get on the bus!" yelled out Hope the supervisor of the day shift.

"Bye Maria, your story is so good and I can't wait until tomorrow to hear some more," said Alvie, the Crochet Lady's best friend.

On the way home from TLC Juanita, a client who had a severe case of Alzheimer's decided to take her seat belt off when Mike was driving her home. She started wandering inside the bus. He had to pull over and in a matter of seconds Juanita went forward and crashed into the front windshield and cut her forehead. There was blood all over the seats, the floor and on some of the clients. Mike kept his cool and made sure that she was okay. Some of the clients were crying and fearful that she would die. In a matter of seconds the ambulance arrived and took her to the hospital.

When I got home I went directly to my bed and slept awhile before supper. I lay in bed and prayed for Juanita. It seemed like the right thing to do. I loved to pray and talk to Jesus. It was so beautiful to just meditate and not only pray for Juanita's healing but also for all of those people who were starving in this world.

Then the Crochet Lady woke up, picked up her crochet bag and started her normal ritual before a session of crocheting. Stretching her fingers, flexing her wrists she commences working on her masterpieces. All of this work to sell to Alvie her best friend at the center.

"Louise, what do you think of this baby blanket I just made for Alvie's great granddaughter," asked the Crochet lady.

"Mom, it's beautiful," responded Louise.

"How much are you charging her?"

"What do you think $5?" asked the Crochet Lady.

"That's too little mom, I'd say probably $10," answered Louise.

"Well it's too late, I told her $5 because she's my best friend," responded the Crochet lady.

She continued to work on the blanket. She finished it making a work of art, like usual. Then she folded the blanket and placed it into her crochet bag. Then she prepared for the next day by loading up her walker with two large bags. She also included smaller pieces of crochet to be sold to the other clients at the center. The Crochet Lady had a little business going on at the center until one of the clients refused to pay her for crochet. Then the center made a policy where she couldn't sell anymore.

It wasn't too much later after that when the Crochet Lady stopped crocheting. Remember the green doily? Well she couldn't remember the way the pattern went. She'd crochet a little, tear it apart, crochet a little and then finally just quit. Her short term memory was dwindling and her finger joints were in excruciating pain from all the years of hard work. But the Crochet Lady with all what was deep inside of her, wouldn't quit and continued with her arts and crafts. Painted pet rocks and cut and paste pictures were among her favorites. She loved to watch "Wheel of Fortune," her favorite game show. Now without the hearing she would just glance at the television and turn around at the kitchen table and continue with her crafts. She worked diligently and always kept herself busy. "Idleness" was not in her vocabulary.

Every morning the Crochet Lady would be awaken by Desa Rey her granddaughter. It was necessary to wake her early because it took her a long time to get washed, clothed and prepared before breakfast. She never asked for help and in her own mind didn't want to be a burden to anyone else. Sometimes she'd want to wear the same clothing for days, which really upset Louise. To remedy this Desa Rey would place a new set of clothing on her bed in preparation for the next day. It wasn't that she was messy, but slowly she was losing her eyesight. The Crochet Lady had an extreme amount of pride and dignity, not in a selfish way but in a way where she held her self in high esteem.

Her husband died in 1995. I asked her if she'd ever get married again and she answered, "What for, so my stepchildren would hate me."

Her response came from seeing all of the people in her generation getting remarried and having family issues with their new step children.

After breakfast she'd wait patiently in her recliner by the front door in anticipation for Mike to pick her up. She sat attentively by the front door because she lost her hearing and couldn't hear the doorbell or the knocking at the door. We gave permission for Mike to peek into the house to see if the Crochet Lady was ready. She then would get up slowly and use the walker to get out the door and walk over to the bus with the other clients.

When she got to the center she immediately looked for Alvie her best friend. There was no Alvie in sight. Then she asked Hope the supervisor about Alvie's absence.

"Where's Alvie? Where's Alvie?"

Maria, Alvie's family has put her into a nursing home because over the weekend, she had a stroke. I don't think she will be coming back for a while.

Alvie was the Crochet Lady's best friend and now there would be a void at the center and in the Crochet Lady's heart.

That afternoon at home the Crochet Lady was subdued in her bed, and very depressed. When it came time for supper she just stayed in bed and refused to get up. Then from the front room we could hear her chanting out loud prayers and petitions to the Lord. She was mentioning prayers for people from the past; her mother, father, her twins, her sisters and her son.

Louise and I were very patient with her and only knew that her time was drawing near.

CHAPTER FIFTEEN

The Shepherd

Pantalion was the first to wake up the next morning. It was always a must to get the coffee going first and foremost. The left over tortillas and fried eggs were eaten alongside green chili. This morning was going to be a time to take inventory of how many sheep were present. Pantalion set out and counted the flock and realized he was short two. Pokey seemed excited. He knew exactly where the sheep wandered off to. He stood up on his haunches and then pointed his nose in the direction toward the cliff. Pantalion could see pieces of sheep wool scattered. He was very disheartened because he knew that Mr. McClellan held him accountable for any lost sheep. So with some caution he looked down below to the bottom of the cliff. He was then sickened to see the remains of an eaten lamb. Pokey seemed to be extra excited and led Pantalion away from the cliff to the path where they came from. There on a beaten path Pokey led him to the sheep's carcass. Pantalion murmured to himself that we better find that last sheep. He was exhausted from the search on foot so he went back to the camp to get his horse.

"*Que paso primo*, (what happened) cousin?" asked Augustine.

"One of our sheep was eaten by a bear or who knows it might have been 'Bigfoot.' There's still one more sheep out there. I'll take the horse out there to cover ground more quickly. I hope the one's okay," responded Pantalion.

So in quest he set out to find the lost sheep. He searched and searched for hours. Finally, Pokey was able to get the scent of the lost lamb. Pantalion yelled out. Then from a distance Pantalion heard a faint cry. Ba ba ba ba, the sound was faint and as he walked closer it grew louder and louder. Long and behold, it was the little critter wedged between two fallen branches. Pantalion was filled with joy. He dismounted off his horse. He then lifted the lamb gently onto the saddle of the horse.

Surely he was the shepherd of his flock, caring for one. It reminded me of how Jesus cared for his flock. When one would go astray he went out to find them and nurtured his lost sheep. It was only the sound of his shepherd's voice, Pantalion's voice, when the lamb responded. With confidence Pantalion was an experienced Shepherd. Sitting tall in the saddle with the sheep was proof that he had graduated to 'Shepherd hood.' From that point on Augustine looked up to Pantalion as the boss.

"Great job, Pantalion, but now we have a problem, how are we going to protect the rest of the flock from the beast?"

CHAPTER SIXTEEN

Life of Boredom

Back at the sawmill...

I was now 21 years old and still without a ring around my finger. Surely enough, my life had to be the most boring life of all boring lives. We went to dances but I was never interested in anyone because my Mount Blanca was coming soon. The Lucero Logging Company was thriving and all the money Tio was getting wasn't coming our way. Catherine, Orelia, and I worked like men. Our youngest sister Emily was still very young and still was growing and getting stronger. Emily was a tough young lady who learned about the mill quickly. She would do all of the odd jobs at the mill like clean ups and carrying the small planks to the truck ready to go off to the Lumber Company. She was like Tio's main worker. She loved to go places with her daddy.

We loaded and unloaded huge logs to be sawed into planks. Tio had a deadline to fulfill for Mr. Nelson at the Vista lumberyard. Every time there was a deadline he would be stressed. So we worked extra hard as we ran the horses to the mountaintop, cut the timbers as low as possible, loaded the timber, sled the logs down to the mill, and then sawed logs to specific lengths.

"Come on girls you can work harder," shouted Tio.

The Crochet Lady

That's all I could hear, "work harder! work harder!"

Sometimes I thought that I was going to lose my mind. I just thought to myself, "Tio was working us like slaves." Just because he doesn't have any boys of his very own he's trying to turn us into men. A man would never be interested in me because my skin was sunburned like leather and my hands were rough like sandpaper. Just look at these hands. Tio never allows us to go into town to do girl things, like shop for pretty dresses and buy something as simple as lotion and nail polish… It felt like I was on a plantation working for Master Petro. All I want to do is sing the blues.

One day my mother decided to chip in and help at the mill. Everyone gave her a hard time and thought that she was out of place. She needed to be in the house preparing lunch and doing the duties of a housewife. As practiced in those days. She just ignored us and pitched in just like everyone else. She didn't hesitate. She was in there lifting and unloading just like the crew. Surprisingly, she was a very strong person. We were all so intrigued by how hard she worked.

"*Vieja*," (old lady) that's enough go get Catherine out of bed. You can go in the house to babysit Paula and Margaret, Its time that she earned her keep," demanded Petro.

At this time Catherine had two daughters who were very, very young and needed close attention. Petro was stressed to meet the deadline given by Mr. Nelson. Catherine came out to take my mother's place on the assembly line. When, in a matter of seconds Constancia turned and got her dress caught on the conveyor belt that was carrying the logs to the big power saw. Logs were rolling to the ground. Everyone was screaming. Her dress was pulling her into the direction of the big saw blade.

It was total bedlam; yes great confusion entered L'Ojito and with confusion came panic.

Everyone dropped what they were doing to help their mom. Petro was livid, cursing up a storm.

"*Pendeja*, (stupid) you know that you are not supposed to wear long dresses on the job, but no you insisted on it."

Petro immediately pulled out his knife and ripped the dress free with one slash and turned to Catherine and physically punished her.

"I'm sorry dad, I'm so sorry; please don't hit me anymore, please. I'm a married woman and the hitting has to stop," pleaded Catherine as she embraced her mother.

"Now my production is slowed down by this "bullshit!" said Tio with frustration."

From a distance was a bright red truck coming down the road. The timing couldn't have been more perfect. Guess who it was? Yes, it was Johnny in a brand new red shiny pickup truck. For the past year he saved enough money to purchase a truck and enough money to move his family out of L'Ojito.

Johnny knew that there was a fiasco there at the saw mill. When he got out of the truck he could see his wife on the ground crying. Then all hell broke loose.

"No man hits my wife!" shouted Johnny

"She's my daughter!" and has been living off me for the last two years. We haven't seen you in all that time. You don't even know your daughters! So get off my property and quick, before I call the law or shoot your ass." Johnny reacted quickly and said, "Catherine!" pack up the kids; we are moving out of this "hell hole!"

He then pulled out his rifle in anticipation of a gun fight. Catherine ran into the big house to get the girls and her belongings. From a distance in the cabin next to Catherine, Johnny could hear Petro cussing up a storm and said, "I'll kill him. I'll kill that "son of a bitch!"

Hurriedly the small rejected family jumped into the car. As they pulled away, Joaquin could hear a ricochet sound on the back tail gate of his new truck.

"What the…," murmured Johnny,"

"So he wants to make it a gunfight?"

It was Petro shooting his rifle at Johnny while Catherine and the girls were inside the truck. Johnny jumped out of his truck, pulled out his rifle and fired back. He then jumped back again into the truck and assured safety to his family as they traveled north. Catherine was so happy to see Johnny. He had just protected his family. This made her so proud to be his wife.

But it seemed like Johnny was a very restless man who sought opportunities to enhance his livelihood. He worked spells in the San Luis Valley, Northern Colorado and New Mexico.

It wasn't too much longer, maybe a year or so later when Johnny and Catherine returned to L'Ojito. Amazingly, on their returned the two men buried the hatchet and reconciled their differences. Petro, although a very tough and shrewd man, he could sometimes show little kindness. When Catherine and Johnny came back to L'Ojito there was an addition to the family, another girl. Her name was Juanita. Petro and Constancia were so happy to see their grandchildren he invited the whole community at L'Ojito. To celebrate this occasion Petro had a great fiesta for their return.

Petro had a tradition. When he invited people to celebrate he'd conjure up his own home made stew. He had a huge black pot. It looked like a witch's pot that would be put over an open fire. The pot was filled with every vegetable and meat you could think of. Rabbit, deer, elk, beef, potatoes, corn, wild spinach, pork, and many herbs and spices were its main ingredients. Every once in a while he'd throw in some mystery meat. Who knows, maybe squirrel meat or opossum. Well, whatever it was. It was so delicious. It was quite a sight to see. All of the neighbors would bring all sorts of side dishes, like empanadas, pies, cakes, tamales, tortillas, and tortas to mention a few.

Even though there was this little excitement for a small part of my life, I was still bored to death and could only envision my man, Mount Blanca looking over that hill with his arms opened for my total love and commitment. Lord, is this just wishful thinking or a waste of my time?

The next day after the fiesta *the Crochet Lady* went about her business and prepared for the day's work, taking out her gloves, Levi jeans, work boots and a sweat shirt. Emily loved to tease her and was told constantly to get off her bed. Emily loved to tease her big sister. But *the Crochet Lady* was tenacious in reprimanding her tough sister who also loved to physically fight with her other sister Orelia.

"When are you going to get married big sister?" taunted Emily.

"It looks like you'll be an old maid, working at the mill until you die." Why don't you have a boyfriend? Asked Emily as she antagonized, the Crochet Lady

"Get off my bed you, brat!"

"Go and take a ride in the truck with Tio. You are getting on my nerves." Emily then left the room and from a distance you could hear Emily and Orelia bickering about whose dress she was going to wear to Mass on Sunday.

Surely, my life was the most boring life of all boring lives.

CHAPTER SEVENTEEN

Once Lost

Pantalion and Augustine were near the end of their journey. The end point brought them through the mountains near Capulin, Colorado. There were 198 sheep still intact and still no French gold.

"It looks like it's time to get on back to Chama. We have had a great season. The sheep are wooly and fat for market and I'm ready to get paid and have a good time," assured Pantalion.

They traveled back south and set up camp about five miles from Chama. They were so proud of their accomplishments but only dreamed about the lost French Gold.

"Primo, let's take one more look at that map and maybe we can drive the sheep in the direction of where the gold might be."

"I think it's worth one more shot and andale pues, (let's do this!)"

It must have been dawn when Augustine and Pantalion were awakened by the barking of Pokey. From a distance Pantalion could see a black bear drudging toward the flock. There was some unsettling amongst the sheep but Pantalion's experience told him not to create a big ruckus. He then saddled up his horse and grabbed his rifle. Then in hot pursuit he quietly took the steed through the flock and then to the meadow where the horse reared up and lifted itself to its haunches. The bear stood straight up to get a smell and a view of the flock. Pokey, then in burning hot pursuit ran directly in a beeline to chase off the beast. Calmly, Pantalion took the horse to a slow gallop. The horse seemed very apprehensive in the chase and Pantalion couldn't figure what was wrong with his horse. So he dismounted and went by foot to see what was wrong. He must have been a mile or so away from the camp and could only hear the faint sound of Pokey's barking. Pantalion was very concerned and shot his rifle into the air twice. Two times was the signal to Augustine that there was a grave problem. The horse like a mule was too stubborn to move so Pantalion as brave as the *Lion's gate* walked into the forest to investigate. He must have walked a mile or so before he could hear Pokey whimpering. He had no clue where the whimpering was coming from. He walked in circles and eventually heard Pokey howling in great pain. Therefore, he found an opening in the ground surrounded with brush. There he found a hole big enough for a bear to make a den. Without a flash light to look into the opening he went into his pocket, pulled out a match, lit it and could see the back leg of his best companion laying there in great pain. He pulled the dog out of the hole and long and behold, there it was a big bar of gold. "Could this be it," he thought? The experience was bitter sweet with the bear attack on his loyal companion and the finding of the gold bullion. Pantalion didn't know whether to scream for joy or cry for his loyal friend. Pokey found the gold, yes Pokey.

Pantalion strained hard to carry the extremely heavy bar and Pokey at the same time. Remember, one of his arms had atrophied at a young age from not getting it set with a cast as a young boy when it was broken. This task was impossible. Hence, he decided to leave the gold in a safe hidden place where he could easily go back and find it. He marked off the area distinctly to trace his way back to get his horse to carry the gold.

By the time Pantalion got back to camp Augustine couldn't believe his eyes. There was Pantalion, carrying the dog. Pantalion's eyes had welled up. He had tears rolling down his cheeks and slowly with a trembling lower lip told Augustine what had happened. Pokey's little body stiffened and Pantalion knew that his dog was dead. Both Augustine and Pantalion sobbed out loudly for the little dog who meant so much to them. Pokey was like family.

"We need to bury this little guy; he's been such a great sheep dog. We will never be able to replace him. He was a family member for sure. So get your shovel out and dig a hole, say a prayer and tomorrow I'll go back to find the treasure."

It was twilight by then and the young men thought about what they were going to do with all the money they were going to make with the found gold.

"Ideally I'd like to own my own ranch and sell sheep like Mr. McClellan. Forget it I won't have to herd anymore and just settle in and enjoy my retirement in Colorado," dreamed Augustine.

"I would love to find a wife and raise a family. I would have to live where it's not too cold and not too hot. I heard that Denver is a nice place to raise that family. Eventually, I'd want to go back to El Guique and introduce my new family to my mother, brothers and sisters. It's been so long ago. I surely miss them all dearly and if Enrique wants to end my life, so be it. Now that I'm a man with money I'd stand up to him and show him that I turned out to be a good productive man. I am not afraid to face him!" Exclaimed Pantalion

Deep into the night the two talked about what the future would bring them with the found discovery. It was truly a miracle.

The next morning Pantalion set out to trace his steps to the gold. He saddled up his horse and felt good and confident that he would secure the gold bullion. He knew that he'd have to travel in a northerly direction and forage through the forest to his marks where he could find the treasure. Visibly, it was placed on a tree stump with tree branches placed over it. Intently, he focused and walked slowly to recall his path and markings. He knew that by horse it should only take a half an hour to find. Oddly, the area where he was tracking seemed so foreign to him.

He thought, "I remember two tall aspens and three pines and then a stump in and near the second pine." "Now that looks like it," but it wasn't.

"WHERE IS IT?" He screamed.

"WHERE IS IT?!"

"DAMN IT!"

"DAMN IT!!!"

He was so frustrated and disappointed, the search was empty. His dream was only a pipe dream and remembered that the previous stories he'd heard about the French Gold where people finding the gold would mysteriously lose the gold. By this time he felt a cold streak down his spine and couldn't get back to camp quickly enough.

To his disappointment he had to tell Augustine that the search was a failure. Augustine was skeptical about Pantalion's response and had his doubts about Pantalion's story.

"Well where's the gold Mr. Denver with the high hopes? Are you shitting me? Are you pulling my leg? Questioned Augustine

"No shitting here primo, I would never do this to you, you have done so much for me, now I believe that the gold is haunted, explained Pantalion. Now, not only have I lost my dog, but now I've lost my future and your trust."

It became a very sad time for the two as they finished the trip and talked very little on the way back to Chama. They were paid generously for the trip in search of the French gold and were given some news about their next employment.

"Pantalion and Augustine, you two are my favorite employees and I do want to give you two, an opportunity of a life time. You see I'm starting a new business in a little town called Capulin. You two will run my ranch and tend to the sheep and cattle I have there. There are chickens and two milking cows which should feed your family. It's a nice quiet town where you'll enjoy your time. When it comes time to graze you will take the flock west," informed Mr. McClellan.

CHAPTER EIGHTEEN

Capulin, Colorado

Moving is never an easy task, but the Muniz family and Pantalion made their way north into a new environment to make a living. Angelita Muniz was so excited about the move.

"It would be a great place to raise our children."

"You know Augustine it's about time for me to have my own senorita. This life of celibacy is for the priests who have vowed to be chaste. But for me I need a woman to love me and give me some great children to love and care for. Just like you Tino, can I call you Tino?"

"Sure "Panty," teased Augustine.

"You know that the last guy who called me Panty got his ass kicked at the orphanage. You remember when I told you the story about Billy Caldwell?"
"But I'm not Billy Caldwell. I'm Augustine Muniz that's who, the man

that's going to kick your ass!"

So the two engaged in a tussle, brawled, wrestled and after the two were too exhausted to continue they stopped.

"You're pretty tough for only being able to use one arm."

"I had a lot of practice dodging the punches from Enrique and the days I was lost in Taos. Do you realize primo; I was actually eating out of trash cans and drinking water from puddles? I had to be a tough sonamabitch. I was like a rabid animal and only could depend on my faith to survive."

"Damn cousin, it is really sad to hear about such cruelty. No one deserves to be treated like an animal," said Augustine with some consolation. "My father was tough on me but always showed us love."

"Well that something I will never get to experience," Pantalion said.

Pantalion then turned away in silence and then in a burst of expression cried out and showed such great emotion that Augustine couldn't help but take out his handkerchief to wipe his own tears.

After sobbing for quite a while, Pantalion said, "I need a wife so I can love my children the way dads are supposed to. You are a true example of a father to me and I appreciate everything that you have done for me."

"*No te preocupes por eso,* (don't worry) cousin, I'll find you a wife. That's the least I can do for you."

CHAPTER NINETEEN

Headaches, Neck Pains, Falls

I'm exhausted. I've shared quite a bit about my life and now it's time to rest. My medicine seems like it's not working like before. It feels like I can't go a day without big doses of "pain reliever." I've been to Chiropractors and have also tried acupuncture. All of the complementary forms of medicine work for a while but a few days later the pain worsens.

Each day, the Crochet Lady's body pain worsens. She sits at the breakfast table in agony; moaning and groaning. She's thinking about the time when she will leave this earth…Her shoveling of oatmeal to her mouth is so very slow…one bite at a time. A sigh… another bite in slow motion… The people who were once from her past, visit her in dreams. She then has conversations with dead people, her mother, her sister Catherine, husband, son, and twin daughters. She chants prayers out loud which have a rhythm in beat. It sounds like singing to the beat of a pow wow drum. Then, she'd screamed out loud. The pain was so unbearable. It was so sad to see her in such a condition.

She calls out to her mother, when in actuality it's her daughter, Louise. Now the roles have changed, daughter has become mother and mother daughter.

"Mom! Mom! I hurt my knee! *Hay mucho dolor*, Pain! Please help me Mom!'

Then Louise lays her ever so gently into her bed. The Crochet Lady screamed for mercy one more time. Then fast to sleep.

She is a very strong woman. She never quits as she would wake up the next morning and take on the day. She starts the routine all over and made her breakfast and waited for Mike by the door to go to the center.

But the next day was different. Mike comes to pick her up to go to the center. Then and as she arose from her seat she glances over toward the bedroom as if she had to go back to her bedroom for something. In a split second she falls backwards like a dive bomber at its target. It is a terrible fall but in a matter of seconds she bounces up as if nothing had happened. It's amazing how she tries to make it look like nothing has happened. Never in my life have I ever seen anyone fall that hard on their head.

Remarkably, the Crochet Lady went to the center that day and was checked by the doctor and checked out ok. The doctor said it was a contusion and her daily medication would take care of it. But it was the beginning of the end. After that day she was having nightmares and episodes of forgetfulness.

Several months earlier, her pacemaker that had a defibrillator attached to it, go off. The force from the defibrillator knocked her to the floor. She screamed out loudly. Her grandson Javier and Leroy a friend who are both pretty big guys picked her up. Then a second shock hit her. The two big guys felt the force from the shock. It was so terrifying to see her in such excruciating pain. In minutes the ambulance arrived and took her to the hospital. The shocks could be deadlier than a heart attack. Weeks later, the defibrillator and pacemaker were turned off. So as a family we decided to execute DNR, do not resuscitate. The simple use of force itself to administer CPR would crush her ribs. In essence, at age 92 we decided to go with her wishes as she decided she would go with God naturally.

CHAPTER TWENTY

Love at First Sight

Pantalion and Augustine loved their new employment on the McClellan ranch in Capulin, Colorado. Raising sheep and fattening them up for market was their trade and expertise. It was getting close to the time of the year to take the new flock out to the grazing pastures. Augustine instructed Pantalion to take the horse and ride westward to scope out areas for grazing. Pantalion was uncertain and was somewhat hesitant because this area was new and foreign to him. Mr. McClellan recommended that area west for grazing and that's where they'd go.

As he started out to the west, the terrain was somewhat dry and unfertile. It was desert like in some areas. Desert like with bunches of sage brushes and cacti along the way. In the distance, he could see the beautiful *sierra* (mountains) getting closer and closer. He stops for a moment. He can't believe the beauty of the mountains. The splendor of the mountains is nothing new, for he's been around the Rocky Mountains all of his life. But, this moment was extra special. The trees were enormous and abundant with every shade of green possible. The fresh scent of the pine trees filled the air. He then took out his binoculars and could see that the area was filled with beautiful green meadows for pasture land. He thought this is the perfect land to graze the sheep. They'll get nice and plump. Pantalion then took out a cigarette sheet and a tin of tobacco. With his good hand he rolled a perfect cigarette. Life couldn't get any better. Or could it?

It was the life he dreamed about, being free from all the pain from all those past years. A tear rolled down his cheek as he thought of his family back in El Guique. My brothers and sisters are probably old now and have families of their own. Emilio, Samuel, Rosa, Dela and Barbara, he thought. They are not the same people as I knew them when I left to the orphanage. My mother is nearly fifty now and so is Enrique. Oh, how I wish I could have changed things.

Over one hill and then over two, then something in the air reminded me of freshly cut wood. I heard a loud buzzing sound in the distance. "What could that noise be?" The noise drew me to it. My horse was uneasy and reared his head. He did not like the deafening cuts of the giant saw blade. It was an interesting sight to see. There it was a logging company out in the middle of nowhere. I had my sights on the workers. Then the more I looked, people were in the form of women. They had the traditional garb on Levis, boots, leather gloves, huge hats, etc. I couldn't believe my eyes!

I got off my horse and stood there for at least an hour. No one knew that I was looking at these incredible women, working like men. They were lifting huge logs at least two feet in diameter and feeding them into the large saw blade. I noticed a man who was directing all of the work, who I assumed was their father. The father was shouting at the top of his lungs trying to push the action of their labor. The father turned to direct one of his daughters to get some water as they stopped for a break and noticed that I was watching these young "lumber jacks" in action. He did not like me looking at them and signaled me to leave. Then the girls turned towards the hill and just stared in amazement. I didn't want any problems so I rode down to introduce myself. I approached the father of the crew and told him that I would be bringing sheep in to his land and would pay him handsomely to graze his flock in his pastures.

"Hello sir, my name is Pantalion Sisneros I work for the McClellan sheep ranch in Capulin. I would like to lease your pastureland in the upcoming weeks. Mr. McClellan said the people in this part of the country usually will talk business."

As Pantalion talked business with this stern entrepreneur, the girls just stood there and gawked at Pantalion. Pantalion felt kind of funny inside and concluded with a firm hand shake and said, "we'll see you in a couple of weeks."

As Pantalion finished the negotiations with Petro, he mounted his horse. Then Pantalion caught the eye of one of the girls. He couldn't stop looking at her. He thought, "Could this be the one I've been dreaming about for the past 10 years?"

"I don't know her name."

Pantalion maneuvered the horse around and went back and approached Petro and asked, "Could you please introduce me to your family?

Petro responded, "Pantalion, you come here with a business proposition and we sealed the deal. This is not a social visit. We have too much work to do here, not today, *caballero*."

As he exited the logging company he heard the girls giggling in the background. "Could these girls be laughing at me?" "Damn! The oldest one was so beautiful." I'll talk to Augustine about the experience I had today. Augustine always has great advice."

Petro was rude to the caballero but it didn't matter because he saw the wife he dreamt of and the future mother of his children.

When Pantalion was about 100 yards away, the entrepreneur whistled and waived him back to talk. He turned the horse around and approached the logger, Petro said, "Oh by the way these are girls are my daughters, Marie Ida, Orelia, and Emily. Catherine my second to the oldest has moved to Greeley, up north. I have three granddaughters too and her husband, Johnny. Also my wife is Constancia. She's in the kitchen preparing dinner.

"Mucho gusto, again, my name is Pantalion Andres Sisneros. My American name is Andy. Well that's what Mr. McClellan calls me." He replied.

"Would you like to come in for a bite to eat?" Asked Petro

"No thank you. My primo is waiting for me back at the ranch in Capulin." During this cordial visit the eldest daughter couldn't keep her eyes off the caballero from El Guique. Like the girl named Marie Ida, Pantalion too caught direct eye contact with her and it was like magic.

Damn, she must have been the most beautiful woman created by God. I already forgot her name. How stupid of me to forget. So for the time, I'll call her Angel.

CHAPTER TWENTY-ONE

White Mountain

The hour horse ride back to Capulin usually takes about an hour. After the encounter with the Lucero daughters it seemed like fifteen minutes. Pantalion was riding on cloud nine.

Back at the center...

"Gee Maria was that your future husband?" asked Hope the daytime supervisor.

I was twenty two years old when this man with a short arm, reddish brown hair and blueish green eyes came into our home. I was old enough to get married five years earlier. But, I waited and waited for my White Mountain. Could he be the one?

It was time to go into the big house and help mom set the table for supper. There was complete silence at the dinner table and Emily and Oralia kept staring at the Crochet Lady waiting for her to say something. They knew that she was touched by the stranger's visit.

"Maria Ida what did you think of the handsome red headed man?" asked Orelia.

"Not much, he's just an ordinary red headed man with a short arm," replied the Crochet Lady.

"What man," asked Constancia as she started gathering the dishes?

"Yea mom, a man wanted to bring sheep on our land and graze them and is willing to pay dad for leasing the land, said a young Emily. Dad, what was his name?"

"His name is none of your business, from Chihuahua, Mexico. Its business talk and like I said many times before I make all of the decisions around here and when I'm not around it's your mother who makes the decisions."

"Come on dad, his name was Pantalion Sisneros from Capulin. We are all part owners in this business and need to learn the ins and outs of this business. I heard him introduce himself. He was a gentleman and very cordial. I'm already impressed with this man… He might be a great catch for Maria Ida," pointed out Orelia.

The Crochet Lady went back to her room after the dishes were washed and laid in her bed and could just see the image of this poor shepherd's face.

"Maria Ida and Emily, you need to go to the stream to get water for tomorrow!" shouted Constancia.

"Mom, why is it always Emily and I who do all of the hard lifting jobs," asked a frustrated Crochet Lady? It seems like Orelia does all of the easy chores. I'm so frustrated, it's time for me to leave this place and start my own life!"

"We all do chores here along with the hard lumberjacking. Today you two carry water pails, and Orelia mops the floor. Tomorrow, Emily mops floors and Maria Ida will go with me to Mass in Capulin to pray for the peace against Germany during Hitler's reign," directed Constancia.

"Mom, I'm 22 years old and without a man. Catherine is 19 and has three children. When will my day come when my Mount Blanca will come and rescue me from this life of work! work! work! I'm tired of it! Some people call me an old maid and I call it, "being very particular."

The two sat quietly as they drove to Mass in Capulin. The Crochet Lady had a tear in her eye which was a rare thing for she was a very strong woman and didn't show very much emotion.

"Well, my hita your day will come. I will pray a Novena for nine days to St. Anthony of Padua and after nine days there will be a sign. By the way who is Mount Blanca?"

"Don't you remember the Chief that came into my school and talked about the legends of the San Luis Valley? I was in first grade when the Chief told the story of Mount Banca and White Butterfly. Remember how interested you were in my story?"

"Now it's coming back to me, yes that was such a beautiful story and I do remember crying."

As we entered the Church I noticed an engraving and figurine on the gate. It was lion's head and an insignia that read, The Lion's Gate, Protector of the Heavens. It was a beautiful fence and one that I never noticed before. Lions Gate, hmm, what can this mean?

At the Masses' devotions the priest asked for special prayers and requests. My mother embarrassed me by asking the parishioners to pray for me so that I'd find a husband. How could she do this? I was so upset with her and didn't talk to her on the ride home. Now everyone knows that I'm an old maid.

CHAPTER TWENTY-TWO

The Conversation

It seemed as if only women attended the Masses in Capulin. Some men would drop their wives off at the front door of the church and then go to the bar to have a couple of beers. When the hour was over the men would pick their wives up and then they would drive home. Most of the people of the area were Roman Catholics and usually kept the day of the Lord Holy on Sundays but The Crochet Lady's family had this different belief. They believed that if you went to Mass on Sundays you could go back home put on the lumberjack garb and off to the mountain top cutting trees. Some of the families in the area felt that was wrong and you know in a small community people gossip.

"Augustine, Augustine, why don't you go to Mass with me? Don't you believe in God?" asked Angelica his wife.

"What a woman, why do we have this same conversation every Sunday, when you come back from Mass? You, my love are the pillar of faith in this family. You are a saint, we depend on all your prayers and I am the salt of the earth and I work hard to support this family!"

"Too many excuses and when it's time to meet your maker what will your excuse be?"

"Just let me be responsible for that because that's between me and God. If I need to pray I'll just go outside to this beautiful valley, Gods country and pray there out loud to God."

"*Mi amor*, (my love) I have great news for Pantalion. You know there's this lady in church who was praying for a husband for her daughter. The girl is really pretty. She would be perfect for Pantalion. Go with me to Mass next Sunday and I'll point her out to you. We must take Pantalion too so he could see her. What do you think?"

"Pantalion will probably go if you ask him. Don't say anything about the girl, it might give him cold feet and you know how shy he is."

"Yes *pero*, (but) he's been asking for a wife, hasn't he? Now isn't the time to be shy, pendejo!"

Then Augustine went next door to have a long talk with Pantalion about going to Mass next Sunday.

"You know cousin, next week is the Feast of La Virgen de Guadalupe and I believe that is would be nice if the entire family goes to Mass on that day. What do you think primo? Afterwards there's going to be a Fiesta to celebrate the day Juan Diego saw her image on his cloak. Who knows you might find an esposa (wife) to start that family you've been looking for," hinted Augustine.

"Do you think so? There was one girl that I saw in L'Ojito, do you think she might be there?"

"Probably, I heard that everyone in Capulin celebrates The Feast of La Virgen de Guadalupe and also the surrounding areas. She has to be there,"

It was a week before the big celebration and Pantalion could not wait to see his princess. He would make excuses why he was scoping out new grazing pasturelands, when he was actually going west to check out the Lucero Logging Company. On a daily basis he would ride north and then turn back west where it was not so conspicuous. But in reality, he went to check out the one girl of his dreams. He found himself a hill that over looked the Lucero property. Without being seen he had a great view of all the action of the lumber business.

He thought to himself, "I like the oldest one. She's so beautiful in her lumberjack outfit. Look at her movement. She moves like a gazelle, so graceful and productive. One day she will be the mother of my children. On Sunday she'll be wearing a dress. What if I don't recognize her?"

It must have been the Friday before the celebration when Pantalion went back to his favorite place on the hill to see his future bride. He found a big rock to sit on. He sat there, day dreaming about this beautiful woman, when he heard some laughter. He then looked down from the hill and could see that the girls were looking upwards. They could see him. Petro wasn't around, which was a good thing. The girls started waiving at him and calling out his name. He immediately jumped on his horse and galloped away.

Pantalion didn't know what to do, so he ran his horse back to Capulin as fast as he could. "They saw me. What did they think of me? I'm so embarrassed."

On his way back to Capulin from a distance he could see a yellow truck coming on the dirt road towards him. When they met on the road he came across the Petro Lucero the owner of the Lucero Logging Company.

"Hey young man what are you doing? Asked Petro.

"Not too much. Just checking out some more pastureland to graze my sheep," Pantalion answered nervously.

"What's your name again?"

"Pantalion Sisneros, I work for Mr. McClellan and presently live with my primo, Augustine Muniz," answered the nervous one.

"Didn't you already check out this area for pastureland?"

"Yes I did but was talking with your neighbors about negotiating prices for the grazing." answered Pantalion sheepishly.

"Okay young man hope you have a nice day. Oh, by the way will you be at the Fiesta of the Virgen de Guadalupe on Sunday? There's supposed to be a big *baile* (dance) after the Mass. I'm going to bring my whole family including my baby granddaughters.

"I'm looking forward to see you and your daughters there on Sunday," answered Pantalion.

Pantalion continued on to Capulin.

"Why did I say, "hope to see your daughters there?"

"Mr. Lucero sure had a puzzled look on his face when I said that. Good thing he didn't see me on the hill overlooking his daughters. Who knows what this tough looking lumberjack would have done to me. I hope the girls don't say anything about me on that hill."

Back at the saw mill...

"Maria Ida, I think that man was really looking at you," teased Oralia as she mimicked two people kissing.

"Now you stop that Oralia!"

The Crochet Lady just turned away, stomped her feet, and walked away. When she turned her back to the girl she had hidden the biggest smile on her face. She then went into her bedroom. She made the sign of the cross and said a prayer, "Lord if this is my White mountain can you give me a sign? If you want me to wait longer, I'm obedient to you. "Thy will be done."

CHAPTER TWENTY-THREE

The Celebration of the Virgen de Guadalupe

Back at the center...

"Maria, why is the Virgin Mary of Guadalupe so important to your people?" asked one of the clients at TLC.

"She is holy. She is the mother of Jesus, who performed a miracle to thousands of the Indigenous people of Mexico and the Southwest. This miracle caused thousands of the Indigenous to convert to Roman Catholicism. Her actual image was imprinted on Juan Diego's cloak. This was way back in the 1500's, where many of the indigenous were believers in many gods. She used Juan Diego an Indian peasant, to perform her miracle at Tepayac, Mexico. She appeared to him on many occasions and ultimately on the last appearance she asked him to collect a bunch of roses and secure the roses in a cloak. When Juan Diego had his cloak full of roses he presented the roses to La Virgen then the cloak opened up with her imprint. The cloak had her imprint. Juan Diego to his amazement couldn't believe his eyes and took it to the local Bishop who also thought it was phony. The Bishop had it test-ed and examined and concluded through many tests that the painting was authentic. The paint used on the cloak was paint not found on this earth."

"Well Maria Ida, we are going to have you teach our religious classes here at TLC," joked Hope

"What happened at the celebration?"

Let me continue the story. That Sunday morning we all got up early, did our chores and got into the truck to make our way to Mass and the dance. We were all looking like good, Sunday girls waiting for a good time after Mass.

I thought about the red headed Mexican who was staring at me from a distance on the hill top. Who knows, he might be at the Mass and dance. I love to dance and I sure hope he does too.

We were all in the back of the pickup and I had to tell Tio to slow down, because the dust was flying all over. My mother was smart; she had us all place a blanket over us so we wouldn't get too dirty.

When we got closer to the church we noticed that the church's parking lot was full. So we had to park in Miguel Garcia's' pasture, next to the Church. When we entered, the usher directed us to our pew, which Tio bought for our family. If you couldn't afford to have your own pew you had to stand up in the back of the church.

When it was time to receive Holy Eucharist I was one of the first to do so. Little by little the people came up to receive the Body and Blood of Jesus Christ. At the very end of the line came up a red headed man with bluish green eyes. It was him. Our eyes locked and before you knew it he was staring at me, yes me. We couldn't stop looking at each other. Is this what love is supposed to feel like?

My mother couldn't believe it. She was telling me in a whispering voice to show respect in the House of the Lord. But it didn't matter I just about broke my neck turning around to see this handsome man walk back to his position in the back of the church. It must have been a sight to see.

At the end of Mass the priest gave the usual announcements and then he said, "Do we have any new parishioners?"

The Muniz family stood up and Augustine introduced his entire family and then looked over at Pantalion and then the priest asked," And may I ask who this is?"

Pantalion slowly stood up and again before he could introduce himself Maria Ida and he locked eyes on each other and there was this long pause and finally, he said, "My name is Pantalion Sisneros."

The entire congregation laughed. Pantalion turned beet red and Maria Ida was also amused by his response.

The Crochet Lady sat back in her lounging chair and said, "I can remember that day like it was yesterday. I just get goose bumps inside of me every time I think of that moment. Oh that feeling of being young again and now I am an old lady who can't hear and walk without help. Sometimes I wish for younger days but then I remember those younger days at the saw mill grinding away."

"Oh Maria, you are so interesting and have so much to offer. You have inspired me to no end," responded Hope as she gave the Crochet Lady a big gigantic hug.

"Hope, I know that my time is coming near. I'm not very productive, I can't help around the house like I use to and when I go home all I do is sleep. I'm getting tired and ready to meet my maker. I just can't stand it any longer." said the Crochet Lady with a cry in her voice.

"Oh Maria Ida, you'll be ok," responded Hope to encourage Maria Ida to survive.

After Mass the entire community gathered at the community center just down around the block from the church. Everyone was responsible to bring a food item and a dessert. The priest provided a goat and two pigs to be roasted. It was quite a feast with a procession from the church to the community grounds. The majority of the people brought frijoles, tortillas and the biggest staple of the area, potatoes.

The fall harvest was the big celebration for Monte Vista down the road. Even after two months after the harvest there were still plenty of potatoes to share with the Capulin community. The people were so content with the meal and then the *musicos* (musicians) arrived with their instruments. Traditionally, there was a guitar, accordion and a violin. The entire dirt floor was filled with dancers, dancing their usual two and four step combinations. Of course there was alcohol served. Many times the men over indulged with the drinking. After the party the wives had a hard time getting their husbands to go home.

There he was Pantalion Sisneros in the corner next to his cousin Augustine. Every once in a while I'd see him looking at me. Oh, how I wanted him to ask me to dance. So the time came when he walked over and asked me to dance. I told him that I needed permission to do so. Petro at first said no and that's where my mom intervened and gave me permission to dance with the dashing young man from El Guique, New Mexico. Oh, how nervous I got. I was in his arms at a length and could actually tell that he was shaking. He never said a word and I couldn't talk either. I didn't know what to say. So I just danced and danced. I just knew that Pantalion was my White Mountain my Mount Blanca.

After our first dance and last dance for the night I said, "Good night White Mountain" and he responded, "good night White Butterfly."

What? White butterfly, I must be just imagining. Did he call me White Butterfly? No that's impossible; he was never in my class listening to the legend of White Mountain. Maybe it was his Spanish accent or something. Oh Lord, are you playing with my mind? Was it that I wanted him to say those words and I just imagined it? Someone once said that the mind plays tricks on you when you are in love.

CHAPTER TWENTY-FOUR

The Letter

Back home at the Muniz residence Pantalion lies in his bunk and ponders about his experience with Maria Ida. "She smells so good and is so soft to the touch. I wonder if she likes me. When we danced I didn't step on her feet. If I would have dipped her or spun her she would have been so impressed by my moves. Petro kept giving me a mean look. but I just looked away. The mother seemed nice, hopefully I appealed to her too. I believe that Maria Ida is going to make a great wife."

The next day Pantalion approaches Augustine with great news. It took Pantalion some time to get the nerve to ask him a very important question. He wants Augustine to write a letter to Petro and Constancia asking for the hand of Maria Ida in marriage.

"Cousin, I have something very important to tell you. I want to marry Maria Ida. I believe that she is the girl of my dreams. She is the one, I know it!" Exclaimed Pantalion

But you just met her, and by custom you need to write a letter to her parents and then they ask her. If they agree they will send you a letter back saying yes and then the arrangements are made for the wedding and if she says no they will send you a pumpkin giving them the sign that she is not interested," explained Augustine.

"I sure hope she says yes."

"Don't worry Pantalion, you'll be just fine."

"I'm 27 years old and not getting much younger. I'm a terrible writer Augustine. Please help me with this letter of acceptance? I do not want to receive a big orange pumpkin. That would really be sad for me."

"Yes, a big pumpkin would be very embarrassing."

"This is hard for me to say Augustine. You know, I've never been with a woman before. Being that you are so experienced, I will need some pointers."

"Now you are getting weird Pantalion. You are making me sound like a Casanova."

So Augustine and Pantalion sat there, rolled two cigarettes and put their heads together to create this letter of all letters.

"You know Primo we need to get your wife Angelita involved. She's a woman; she'll know what parents want to hear."

It was a matter of minutes and before you knew it they created a masterpiece. It was handwritten in Angelita's best penmanship. They used ink and quill with special paper that they got from Alamosa on their last visit. The letter said:

> Dear Mr. and Mrs. Lucero,
>
> It is our pleasure to send you great news. Our cousin Pantalion Sisneros would love to ask your daughter Maria Ida for her hand in Marriage. Pantalion has been saving money for this very special day. We have an additional house here in Capulin for the two can live and work on Mr. McClellan's sheep ranch.
>
> In the next two weeks Pantalion, Angelita and I will visit you to get your answer. If not, a pumpkin is not necessary. We

are adults and can take the bad news like adults. But we are confident that the answer will be yes.

*Yours truly,
Augustine Muniz*

It was the next Sunday after the celebration of La Virgen de Guadalupe, when Constancia received the letter from Angelita. As usual the men went to the cantina when the women went to Mass. The women as routine knew that at the end of Mass their husbands would be their waiting in their proverbial trucks. It was tradition, customary for every man to have a truck. It was a man thing. It was a "Man's world" for sure. The head of the family was the man and in this case Petro was definitely "El Hombre." The buck stopped where he put down his axe or his saw. He only got advice from Constancia on occasion and this might be the time when they received the letter.

Constancia did not share the letter with the Crochet Lady. The letter was written to them from Augustine and when Constancia showed the letter to Petro there was a negative reaction. He was very upset because he knew it was the Crochet Lady's time to leave the nest. How could he let her go? She was his best lumberjack.

Behind closed doors Constancia said, "We just can't tell Maria Ida anything yet.

"Let's give it a couple of days before we tell her," said a disgruntled Petro.

Back at the Muniz home...

"Angelita, did you give La Senora Constancia the letter of acceptance?" asked Pantalion.

"Yes Pantalion, but I did not wait to see her expression."

"We will have to go visit their family next Sunday and wait for their response.

The entire week was purely anxiety for Pantalion and the Muniz family. In discussion they feared that the Crochet Lady would give them a negative response.

Back at the center...

I was truly scared to make the commitment to Pantalion. He was 4 years older than me, but that was what I wasn't afraid of. I'd never been with a man before. Back then couples never went on dates. Sometimes parents just made the arrangement. In some cases one would marry a distant cousin. But not me, it would be good stock and I'd be marrying someone who was not a relative but someone who lived as far away as El Guique. Decisions, decisions, something I never had to make, a decision. It was always mother and Tio who made all the decisions.

The Crochet Lady then set back in her wheel chair and got an excruciating pain in her neck. The group of clients sat there and stared at her for a minute or so. They only hoped that she would continue with her story. The Crochet Lady's story was similar to their own stories on how they met their spouses. You meet someone fall in love, grow old together, and continue into the next world.

The rest of the evening was a very hard time for her. She pretty much lost all strength in her legs and needed us to help her in and out of bed. Also, she needed help in and out of the lounger and in and out of the toilet. She was in great pain and we got news from the TLC agency that she would need a long ramp out the front door to the street so the TLC bus would be able to pick her up and transport her safely. Her pride was huge and even then she tried to walk to the front door. Her pain was immense and when she slept she'd moan and shout out words to people who had passed on before her. In these last days the spirits from the past came to visit.

One day, one of her sisters came to visit her who she hadn't seen for 30 years. Little Emily was now in her 80's and the Crochet Lady did not recognize her. Emily was no longer the little baby sister that she once knew. Emily was so sad to see her sister in such a condition. The Crochet Lady just sat there and stared and smiled.

Time had elapsed and her time was drawing near. Another visitor was her eldest grandson "Little Andy" and his wife Jeanna who sure were a big help in her last days. They comforted us bringing food and great company. They loved their grandma dearly. Her grandchildren all came in to visit, Dan and Diego and their children also, Melody, Carlos and children from Oklahoma, Andy her son from South Dakota, who she hadn't seen for a while, Andy's children from Northern Colorado Lori and Mark and Jody and Rick, and their children Raven and Lila, Desa Rey and Javier, and all of the Martinez clan. It was quite the reunion but not a fun time.

It was time to make the call to the Church. Father Jorge came to our home to give the Crochet Lady her Sacrament of Healing. Some call it the "Last Rites."

Two years earlier was a better year as we celebrated 90 years of her life and two years later she was fading away. One day she had awakened from a two day sleep. She was coherent wide awake and talking up a storm. She knew what was going on around her and concluded the story about Pantalion. It was like she knew that her story needed some closure.

She said, "You know Albert, I made Pantalion wait two years before I would tell him yes?"

"Well grandma, why did it take so long?"

She answered," Because Tio didn't want to let me go. I was his best lumberjack and he needed me to make him money. So the negotiation was that if I would marry Pantalion, Pantalion too would be a lumber jack.

But I wasn't too sure about this man. Was it fair for him to live as a lumberjack, a very hard and enslaving life? Just looking at him I knew that he was kind and probably would make a good husband. But, I feared that Tio Petro would chase him away like he did Johnny.

"Oh what should I do?"

My prayers were deep. I didn't want mom to miss me. Tio Petro insisted that if I would leave that I better not come back begging for money. I think that was the breaking point. I knew that Pantalion with God's blessing could make a beautiful marriage.

CHAPTER TWENTY-FIVE

The Visitation

Finally the three of us, Petro, Mom and I got together. To my surprise I found out that Pantalion wanted to marry me. I was so excited but I contained myself knowing that Petro wasn't too excited. The time was here and the most important decision in my life. Now was this a great idea? Tio kept insisting that it was too quick of a decision to make. My mother was hesitant about the marriage and very protective of her daughter. There were questions that needed to be discussed and answered. Would sheepherding be enough to support us? Where would we live? Tio felt that sheepherding was not enough and questioned why Pantalion was still living with the Muniz family.

Soon Pantalion, Augustine and Angelita would be coming over for their answer. It was time for the Luceros to come up with their own counter to the letter. Would the answer be, yes or no?

We decided that there needed to be a waiting period so we could get a better understanding of who we were. So when the proposal came to our home, we decided that there would be a waiting period, an engagement was set with no date.

When they came to the door Pantalion came with flowers in hand and a bottle of homemade moonshine, which Tio was very interested in. The discussion went on for hours.

"Pantalion, I'm so sorry but I cannot let you marry our daughter. There has to be a waiting period. My daughter has asked for time to ask the Lord if this is the right thing to do and we need to do the same," stated Petro.

There was a sour look on their faces and the conversation switched to normal everyday talk about the families, business and the weather. Pantalion went into complete silence. I could not look into his beautiful blue eyes.

"Petro, thank you for your time and I am willing to wait for my future bride to be. I will not be with her until it's time for the wedding, sir," answered Pantalion.

When the company left L'Ojito, the Crochet Lady walked slowly to her room and lay in her bed and prayed for hours. She asked the Lord if this was the way it was supposed to be.

Her prayer was, Dearest Lord, if this is my White Mountain, I will know by your guidance. I trust in you fully and look to you for everything. You are my salvation and life, amen.

The Crochet Lady at a young twenty three years of age corresponded with Pantalion weekly. At Sunday Mass at Capulin they would pass letters back and forth. Oh! How there was a deep yearning for each another. Except for one day, their message spoke about a rendezvous, a secret meeting in a local cave near the Lucero property. No one knew about the cave except for Emily and the Crochet Lady.

Pantalion wrote:

> *My Dearest Maria Ida,*
>
> *I can hardly wait for that day when I can hold you in my arms and gently press my lips on yours. It has been a long two years but you are worth the wait my love. If we could only meet in secret so I can see you and love you.*
>
> *Truly,*
> *Pantalion a sheepherder*

Ironically, it was Augustine and Angelita who were helping him with his love letter writing. Pantalion never learned how to write well in the orphanage. He was too busy building projects. But anyway, the Crochet Lady was so impressed with the beautiful words which she never heard a man tell her before.

She always hid her love letters from the rest of the family because her two younger sisters loved to tease her. And, what if Petro found out?

The Crochet Lady wanted to see Pantalion so badly. They thought of a way that they could meet in secrecy. She thought and thought and remembered the cave on the hill where they had found choke cherries the year before. It would be the perfect place to rendezvous. Tio would be gone on a business trip to Alamosa and her Mom pretty much stayed in the house all day doing house work. Now, she had to do something to detour her younger sisters Orelia and Emily. She put her brain to work and thought about having the younger siblings work on a puzzle that she had been saving for a time when there was nothing to do. So that was the plan. Now I'll have to write a letter to him and hopefully he will come to my cave. Now she had to conjure up the letter. She wrote:

> My dearest White Shell Mountain,
>
> It's been over a month and how I long to see you. My step father is leaving and going to Monte Vista for business. We can meet in a cave right above our home about a half of a mile to the north. I want to hold you tightly in my arms and share who I am. I will place ribbons on the path so you will know how to get there. Remember next week on Friday at 1pm. I don't expect Tio to be back until Monday. He usually stays all weekend long. I'll make it a picnic. I'll bring the food and you bring the conversation.
>
> Yours always,
> White Butterfly

So the rendezvous was staged. Pantalion received his love letter at Mass that previous Sunday. He would meet her in a cave a half a mile north from their home and it would have many red ribbons leading up to it.

CHAPTER TWENTY-SIX

The Rendezvous

This was a daring move on the Crochet Lady's part. She was daring all right. Her mom stayed in the kitchen making empanadas and the girls were working on a 500 piece puzzle. She thought to herself that should keep them busy.

She asked her mother for a couple of empanadas and when her mother was not looking she sneaked a couple more. It was going to be a great moment and nerve racking to say the least. It was the first time the Crochet Lady would show her deceptive side and be with a man.

Pantalion on the other hand was so nervous, but was able to get enough courage to ride to the rendezvous point. The Muniz' prepped him for this meeting and instructed him to be a gentleman and show total respect for her. He too never experienced being with a woman so it was all new for him.

"Augustine, what will we talk about?"

He responded, "Just tell her what you know about sheepherding and then gradually tell her about your incredible life story. She will be totally interested in the man from "El Guique."

"My story is not incredible, it's sad!" responded Pantalion.

"Now settle down, I know that you are very nervous and yes scared. You will meet your maiden and shine like a champ. The horse is outside saddled up and ready to go. Now say a prayer and *hasta, entonces.* (Bye)

Slowly Pantalion made his way to the little hamlet. The key was not to be seen. It was 12:45 and 15 minutes to the rendezvous place. To get to the cave he took the long way around to L'Ojito from on top of the tree line north. Then he saw one ribbon then two and then three. Yes, Pantalion found the path to the cave. He made a whistle like sound to see if he would get a response. He whistled again and out from behind a bush in front of the cave she came out and greeted him.

"*Hola*, hi," she greeted.

"Hi, (long pause) I'm Pantalion Sisneros, the man who wants to marry you and my life is so sad.

"Don't be so silly Pantalion; of course I know who you are. Come into my cave so we could have some fresh empanadas. Do you like empanadas?"

So after a couple of minutes the two just stared at each other and had nothing to say. It was so quiet that all you could hear was the sound of the birds whistling their song.

Pantalion stood up and said, "Well it's time for me to leave, Augustine needs me back at Capulin to gather the sheep."

He extended his hand and she extended hers and before you know it they were in an embrace. They held each other tightly and both cried because they were so happy. They were tears of joy of course. Then he decided to stay a little longer as he wanted to tell the Crochet Lady about his journey from El Guique to L'Ojito. So the two sat down on the cold cave floor and listened to the story of Pantalion's odyssey.

The Crochet Lady in turn shared with Pantalion that she had lost her father at a young age and how her mother and Tio Petro raised her. She also told him about the story of Mount Blanca and White Butterfly and how both of their own lives reflected its legend.

Pantalion was so intrigued to hear the story from this beautiful lumberjack from the la sierra, L'Ojito. He thought, "is this to be my lady forever?" The story was so true to both of their lives. He was White Mountain and she was White Butterfly his maiden.

"My sisters are my half-sisters, but I love them like they're my real sisters. I had a real brother who ran away from home because he didn't like Petro. Petro was very "strictly" with him and me." She responded.

"It's time for me to go now and I do want to see you again. Can we make this our secret cave that no one knows about?" He asked.

"We'll see. I don't know if Tio is going to Monte again next week but…" At that very moment Pantalion got up and grabbed her hand and held her in his arms and gave her a big kiss that lasted for minutes. She did not refuse him and melted into his arms. Then the two sat back down talked some more, kissed some more until twilight. They gave their goodbyes and after that, they continued to exchange love letters on a weekly basis.

When she got back to the house it was already dark and her mom was worried sick that something bad had happened to her.

"Maria Ida, do you realize that something bad could have happened to you. Where have you been?"

The Crochet Lady responded. "I got lost and had a hard time finding my way back."

The girls were still working on the puzzle and Orelia responded, "Mom she has a funny look on her face, I think its called love. She got lost because she was dreaming about that sheepherder from El Guique. Just look at her eyes, *puro amor*. (pure love)"

"Oh shut up Elsie or shut up Orelia, What is your real name? I don't know how you got named Elsie from Orelia. It sure doesn't sound like Orelia to me, Elsie, Orelia, Elsie, Orelia. See it makes no sense! So, you've never gotten lost before? So shut up!"

In the meantime, Pantalion had a hard time finding his way back to

Capulin. It was so dark he couldn't see his hand in front of his face. He had to depend on his horse's instinct to ride back safely. It really didn't matter because he was in love with his lovely maiden. She would be the one he'd been waiting for all of those years. She was his dream, his White Butterfly.

When Pantalion arrived at the Muniz' home they were all waiting to hear the news about his first encounter with Maria Ida.

"Okay Pantalion, tell us everything about your first kiss," Augustine said with a teasing look in his eyes.

"Shut up primo, it's none of your business!" A red-faced Pantalion said sternly.

"Well tell us something. Don't just stand there!"

"It's like this primos, we had a lot of fun and talked and talked and what seemed like minutes turned into hours," responded Romeo.

"Did you kiss her?" Asked Angelita?

"Don't you think that that's kind of personal?

After two years of waiting, Pantalion and Maria Ida Sisneros were man and wife. They built their own home on the property of L'Ojito and worked awhile at the Lucero Logging Company. After the two couldn't make a living at the Lucero Logging Company they migrated north to Weld County and raised a family and worked as migrant farmworkers. Pantalion was injured in a car accident which scared his fore head and left an indelible horizontal line across his forehead. They had two sons Andrew, Alonzo, two twin daughters Juanita and Rosana who died at a young age and Louise my wife who after years of prayer became the Crochet Lady's baby girl who she'd been praying for to heal her hurting heart after the death of her twins. Pantalion was able to make amends with his father Enrique when his eldest son Andy was first born. It seemed like Enrique had forgotten about everything and buried the proverbial hatchet of hatred.

The Crochet Lady never went back to the center. In fact all of the clients who were so intrigue with her story never got to hear the ending. To this day the clients are waiting for her to come back to bring closure to this beautiful story.

CHAPTER TWENTY-SEVEN

Crossing the Bridge

In her last moments of life, we set up a bed next to hers. This allowed us to monitor everything the people from Hospice advised us to do. Her breathing had been heavy and shallow. You could just see the pain she was experiencing. I was sleeping in the other upstairs room when Louise called me downstairs. We maneuvered the Crochet Lady's body back to her side to make it easier for her to breath. She was without words. We tried to stay awake to be there for her last breaths, but slowly we drifted to a deep sleep. Then I was awakened by a loud gasp for air. It was her last breathe of life. Maria Ida Sisneros, the Crochet Lady died a day before Thanksgiving in 2014. She was a sweet loving lady who always put her children first above all. She always thought of her life as being so boring, but as you witnessed through this reading she was a true contributor to Colorado History's finest. She helped build Colorado with a lot of sweat and hard work. Not too many can say their mother was a lumberjack. So like the Legend of White Mountain, White Butterfly finished her life on the mountain top.

Her life did not end here for her goal was the mountain top. She was a believer of eternity and her aim was the Mountain top to be with her White Mountain who passed on in 1995. As her spirit lifted from her body she best explained her journey.

My pain was released. I had no more pain and felt strong and energetic like a youngster. Everywhere I looked was darkness. I stood in silence for a moment that seemed to last forever. When I looked down I could see my past on earth. There I was in the arms of my mother at the funeral of my father Juan Pablo. Then I could see myself giving birth to my twins. I could see my daughter graduating from college. I could see Alonzo in his greatest need in his fight against cancer. I could see Andy on a roof top carrying bundles of shingles. There was Loric my granddaughter giving the commencement speech at Colorado State University. All the scenes of my past were before my eyes.

The scene changed. Louise and Albert were by my bed side. They were both in grief, sobbing out loudly. I felt so badly for the two and wanted to return to comfort them. Then I could see what looked like my lifeless body being taken away by two men to the funeral home. Then there was this voice calling out my name. I turned toward the voice. The voice then turned to voices. They were singing out my name, Maria, Maria, Maria we love you come, come, come please come.

I thought, "What could this be?" "Could this be when I meet my maker?" Maria, "You must cross the bridge to eternal life," an angelic voice sang out."

All around me were white billowy clouds. When I took my first step forward the clouds slowly lifted. I was able to see what was before me. It was the bridge that I heard my priest talk about at many funerals. It was a beautiful brown bridge that was made from the wood that I use to cut down at L'Ojito. The smell of the fresh cut wood was in the air.

I heard a loud booming voice saying, "Come across Maria. Come meet your new eternal family."

I felt at peace within me, "Could this be judgement day?"

From a short distance I saw what looked like hands extending, many hands reaching out. Voices were in the background, many different languages. The sounds were angelic to my deaf ears. I could hear again, "alleluia!" Every step I took I could felt the strength in my legs. I had no neck pain. I was walking upright. I was running. "Praise the Lord!"

The bridge wobbled and swayed back and forth. I looked back. The bridge was crumbling. I ran as fast as I could. The boards beneath me were dropping into the abyss. Then the hands were reaching out. I recognized a face at the end of the bridge. Wow, could that be my mother? Could this be the reunion of all my people from the past? The bridge stopped swaying and crumbling. The ground was stable. Slowly with every step forward I could see someone who I thought I knew. The picture became clearer. There he was my White Mountain waiting for me with his arms extended. He was whole. He looked like a 20 year old, so very handsome. He had no wrinkles. His scar on his forehead was gone. Most notably my husband's broken arm was like a new healthy arm. He was a new person. When he hugged me it was a full hug tight but gentle. When we kissed it was like when we were in the cave in L'Ojito. I felt younger too! I looked down at my hands. All of my wrinkles were erased. My legs were strong and I didn't have that annoying neck ache. There was a rumbling in the air. It was the sound of people's voices. I could hear everyone talking. I was a new person. There was my beautiful mother. She looked so young. She was very happy to see me. We wept tears of joy. We embraced and when she held me I felt the warmth of her love. It was like I was her infant at the funeral of my daddy, Juan Pablo. There was my daddy Juan Pablo, exactly how mother described him. He was so excited to see me. My brother Regino was there too. My daddy and he looked so much alike. My younger sisters Orelia and Catherine approached me. We held each other tightly. We cried and cried. Our tears were tears of joy, tears of triumph over death.

"Maria, I would like to introduce you to the people who you once knew on the other side of the bridge… You knew them once in your life and at a young age they left you and Pantalion to be with me," said a voice from above.

Then three persons walked forward, her twins Juanita and Rosana and her Alonzo. Juanita and Rosana were all grown up and very beautiful. Maria said, "I missed you so dearly and I prayed that one day I'd see you again. My prayers have been answered. You also have a baby sister. Her name is Louise. You'll love her."

Alonzo then grabbed her and held her in full embrace. He said, "Thanks Ma for taking care of me during those trying days. God had a plan for me to be with him in paradise. I love you Mom."

CHAPTER TWENTY-EIGHT

On the Mountain Top

In my house there are many dwelling places. If there were not, would I have told you that I am going to prepare a place for you? John 14:2 King James Version

This moment is beyond beautiful. Its presence is so lovely and peaceful. There are neither worries nor pains, just true everlasting life with Him.

Then I heard a voice from above say, "Maria I've prepared a place for you."

The Crochet Lady responded, "Thank you Lord, oh, thank you Lord!"

He then grabbed my hand and walked me to the mountain top to see the out reaches of Heaven. I looked up to see His face and saw brightness, but not the kind of brightness that blinds you…The kind of brightness that makes everything clear and vivid. As I looked up I felt his hand touch mine. I felt no fear, because I felt His magnificence. There it was in front of my eyes, His Heavenly Host. As far as my eyes could see were choirs of angels giving him praise. He then took me to my dwelling place. There again I saw all of my relatives rejoicing. I had a special room where all of my crocheting materials were made available to me. The room was filled with yarn of all colors of the rainbow and colors not known on earth.

Then He said, "Crochet for me, crochet."

In amazement I said, "What shall I crochet my God?

He answered, "The green doily, the one you couldn't finish on earth." The Crochet Lady waits patiently for eternity to begin. She sits on her favorite chair by the opening where she sees the Heavenly Host singing praises to God. The rays from His heart are like the sun. They gently hit her face. She feels its warmth and her fingers gradually begin to respond. Exercising her hands she clenches and then quickly opens, clenches and reopens. Then like a conductor of a symphony or like a baseball pitcher warming up before the big game, she gracefully moves the yarn to and fro with her needle and flexible fingers. Thus, following a pattern that will make the most beautiful green doily for God. Pausing for a second, she reaches for a ball of yarn. There was no hesitation for she remembered the pattern so vividly. She then takes in a big breath of Heaven and whispers to herself, "Oh, how time has flown, it was well worth it. My Lord, I've been faithful to you."

THE END

CPSIA information can be obtained
at www.ICGtesting.com
Printed in the USA
BVHW031111091120
592860BV00002B/29